Foot and Mouth Disease

Foot and Mouth Disease

New values, innovative research agenda's and policies

A.J. van der Zijpp

M.J.E. Braker

C.H.A.M. Eilers

H. Kieft

T.A. Vogelzang

S.J. Oosting

EAAP Technical Series No. 5

CIP-data Koninklijke Bibliotheek, Den Haag

ISBN 978-90-76998-27-5
ISSN 1570-7318
paperback

Subject headings:
A-list diseases
Participation
Epidemiology

First published, 2004

Wageningen Academic Publishers
The Netherlands, 2004

Wageningen Academic
P u b l i s h e r s

Table of Contents

Preface

The outbreak of foot and mouth disease in the Netherlands in Spring 2001 became a national crisis. Concern for the livestock industry and rural areas was clearly visible in all actions and reactions to the crisis. These social and scientific expressions have been linked together in the dialogue in the Foot and Mouth Disease (FMD) Workshop. We wanted to use the discussion on the experience of stakeholders and scientists in different disciplines in formulating themes for the research agenda of the Wageningen University and Research Centre (University of Life Sciences). The research agenda should address the issues raised by stakeholders and integrate different scientific expertise. The results should support farmers' organisations in policy development, and the government in the prevention and management of future outbreaks of Foot and Mouth Disease.

The steering committee and participants in the FMD Workshop have learned much from this workshop approach. The sustainability of this approach, however, depends on how the workshop results and recommendations are used. We hope that stakeholders will continue to voice their opinions in setting up and carrying out of future research on Foot and Mouth Disease. Furthermore, we hope that students at the Wageningen University and Research Centre will carry out interdisciplinary studies and research on the issues set down in the research agenda. We can now offer support on policy issues with process experience and in future, on questions arising from research on these issues.

The outcome of the FMD Workshop is the report entitled "Foot and Mouth Disease: new values, innovative research agenda's and policies" and is the property of all participants. A number of colleagues deserve special acknowledgement for their work and vision in setting up, implementing and presenting workshop outcome. Karen Eilers and Marleen Braker took the lead in designing and organising the workshop. We are very grateful to Henk Kieft and Simon Oosting for their support in facilitating the workshop and their efforts in organizing many others. The confidence of the Board of Management of Wageningen University and Research Centre in creating this dialogue between stakeholders in the FMD crisis and scientists illustrates the social importance of the FMD Workshop and through it, of Wageningen University, in underpinning social issues with education and research.

I hope that the FMD Workshop will advance the process of innovation and expand the relevance of Wageningen University in the wider society.

On behalf of all participants and the Animal Production Systems Group, the Department of Animal Sciences,

Akke van der Zijpp

Chapter 1. Values at stake during the 2001 Foot and Mouth Disease epidemic in the Netherlands

In 2001, the Netherlands was stricken by an epidemic of Foot and Mouth Disease (FMD). The epidemic had a great impact on the whole country. Since dairy farming accounts for approximately 70% of the country's agricultural land use, it was dairy farmers who suffered the most serious effects. However, what is remarkable was that the epidemic developed into a crisis affecting more individuals and groups than those involved in that sector. This was not because nobody was prepared for the epidemic; indeed, the Ministry of Agriculture, Nature Management and Fisheries (now the Ministry of Agriculture, Nature and Food Quality) already had an action plan addressing it. This consisted of a scenario specifying the necessary action, tasks, competencies, and responsibilities of various stakeholders in the event of an epidemic and it was implemented as intended. Moreover, the scenario did what it was supposed to do: it prevented the spread of the disease, eradicated it relatively quickly, and minimised the effect on agricultural exports. Nevertheless, there was in fact a crisis.

The epidemic became a crisis because many groups and individuals, including many of the farmers affected, neither understood nor accepted the measures that were imposed by the scenario, in particular the killing of healthy animals. Discussion still continues (see, for example, Cuijpers & Osinga, 2004; RLG/RDA, 2003; LNV, 2003) as to whether healthy animals should be killed during any future FMD epidemic so as to prevent the disease spreading or whether animals should in fact be vaccinated, at least those in zoos or those kept as a hobby animal. The reasons for the controversy will be explained below.

1.1 The production value of land use

Something had apparently changed in Dutch society between the action plan being drawn up and its being implemented. According to a taskforce set up by Wageningen University and Research Centre in 2001 to consider the future of Dutch agriculture (*Taskforce Waardevolle Landbouw*, 2001), the structure of Dutch society in the 1980s and early 1990s permitted an approach to tackling infectious animal diseases such as that set out in the scenario. A decade later, however, the same scenario led to a crisis because it conflicted with the new values that the Dutch associate with the countryside.

As in other countries, agriculture in what is now the Netherlands, developed thousands of years ago as an activity involving the whole population. That remained so for a very long time; even with rising urbanisation, a major proportion of the Dutch population continued to be involved in farming and the countryside. Agriculture was the concern of all; everyone understood the importance of agricultural food products for their survival, and towns and the countryside were not separate as they are today (Bieleman, 1992).

In the Netherlands, specialisation in a small range of products intended primarily for export began centuries ago (Bieleman, 1992). Specialisation and the focus on exports were boosted after the Second World War by the combined effects of policy, research, expansion, and education. The objective became relatively simple, with land in the Netherlands being used to produce a "value" that enjoyed wide acceptance, namely cheap and abundant food (Frouws & Leroy, 2003; Van der Weele et al., 2003) of high quality (Taskforce Waardevolle Landbouw,

2001). Scientific researchers and government authorities were spurred on to support this value by international competition, which demanded continual price reductions and improvements in quality. As a result, supporting values such as institutional and technological control of production processes and products, mass production, and product uniformity became associated with the use of land to produce abundant, cheap, high-quality food (Figure 1).

System values production agriculture	System values transition farming

Low price
Food abundance
Physical quality
 	
Control
- *technological*
- *institutional*
	Sustainability
	Food values
	  Metaphysical quality
	  Authenticity & care
	  Naturalness
	  Confidence and
	  transparency
	Landscape values
	  nature
	  culture history
	  open, silent and quiet
	  accessibility

Figure 1. System values of production oriented and transition agriculture (Adapted from Oosting and De Boer, 2002).

Moreover, key stakeholders involved in land use – farmers, processing industries (which were mainly exporters), and the Ministry of Agriculture – were unanimous in their attitude to most of the relevant issues. Moreover, this "green front" had a monopoly of knowledge and expertise and a great deal of political influence; the rural population provided the backbone of support for the Christian Democrat parties at the centre of the Dutch political spectrum, parties that have helped form the country's ruling coalitions for most of the past 50 years. Most Dutch people were proud of the achievements of the country's food producers and the Netherlands was seen as an example for the rest of the world, indeed as a solution to the evils of starvation and malnutrition in less favoured countries. All in all, land use in the Netherlands was for decades associated with feelings of pride and consensus, feelings expressing themselves in concrete form as a simple objective, namely the production of cheap, abundant, high-quality food products.

1.2 Land use in transition

In the relatively prosperous and urbanised societies of Western Europe, food has become something that is taken for granted. Once basic needs were satisfied, higher ones such as social relationships and self-actualisation became increasingly important (this is in line with Maslow's predictions (1954)). Enjoying the landscape, nature, and culture became important factors in self-actualisation. Recreation, keeping livestock as a hobby, art exhibitions in rural settings, and enjoyment of the beneficial effect on health of a quiet, peaceful, green, and open landscape are just some of the ways people make use of the various non-production functions of the countryside (Van der Ziel, 2003). Some farmers have taken up the challenge of creating a new supply side to meet consumer demand and ecological requirements and there are numerous initiatives in this field (see Van Broekhuizen et al., 1997; Van der Ploeg et al., 2002). They include "agrotourism" and other recreational services, green "social care farms", regional products, organic production, and a variety of landscape and nature conservation schemes. Such initiatives are often found in combination. Moreover, a growing number of countryside dwellers are not involved in farming and if they do keep livestock it is as a hobby rather than for any economic purpose. As a result of all this, non-production values (referred to in Figure 1 as transition values) were widespread in the areas affected by the FMD outbreak.

1.3 Conflicting values

The two sets of values in Figure 1 would seem to conflict because nature and landscape values have fallen victim to rationalised land use. Streams have been straightened, plots of land levelled, and hedgerows removed. Artificial fertilisers, drainage schemes, and irrigation have made agricultural landscapes virtually identical throughout the country, and the excess of nitrogen and phosphorus produced by farms threatens nutrient-poor biotopes and water quality. Moreover, in order to prevent infection, farmers discouraged unnecessary visitors and when people did come into contact with farming or watched it on television, they perceived it as an industrialised activity in which animal welfare standards could not possibly be acceptable. (This was quite apart from reports showing cranes hoisting up animals slaughtered preventively during the FMD epidemic.) Production-oriented farming perceived changes in land use as a threat, while the transition to new types of land use was hampered by the traditional focus on production.

Each set of values had its own interest groups: production values were in the interest of export-dependent processing industries, farmers who were economically dependent on those industries, and elements within the veterinary authorities. Societal, transition values, on the other hand, were cherished by the majority of the population. Moreover, both value clusters sometimes co-existed, even within such stakeholder groups as the Ministry of Agriculture, farming organisations, research institutions, and even individual farms, for example where milk production was combined with recreation, social care work, or keeping hobby animals. Values were therefore a matter of conflict both between and within groups of stakeholders; moreover, many groups and individuals did not realise what values were actually behind their behaviour. Values are after all implicit in the way people think and act and not explicit.

The Foot and Mouth Disease workshop reported on in this book was intended to make individuals and stakeholder groups aware of the values behind their behaviour and to initiate a dialogue between the different groups on a basis of respect and understanding for the context in which each of them operated.

1.4 The workshop

The "Valuable Agriculture" taskforce presented its report *Values for a Valuable Agriculture* (Naar een WaardeNvolle Landbouw) (Taskforce, 2001) on 11 October 2001, with the following recommendations:

Wageningen UR should initiate and facilitate serious discussion of agriculture, recognising that science is not a value-free activity. One's own values should be made explicit and deployed as a force within the process of change. We also believe that Wageningen should pay much more attention to the way in which change takes place. The task force has considered this, but it is still a virtually unexplored field.

Our second recommendation to Wageningen UR has to do with the method referred to as "Sharing knowledge is creating knowledge". Problems and solutions in agriculture cannot be divided up into scientific disciplines and areas of expertise. Collaboration…is of great importance. Solutions to problems should be sought in collaboration with those concerned, with the various different parties within society. That is the only way to bring about true innovation.

One initiative that gives concrete shape to these recommendations at Wageningen UR is that of the Wageningen Workshops. These are study and design workshops at which public debate plays a major role; they deal with specific problems in collaboration with those involved. Wageningen UR organises and supports the workshops with relevant knowledge and expertise. This is by definition an interdisciplinary matter. The topics for the workshops are determined on the basis of public interest.

(Quotation from an address by Wijnie van Eck at the presentation of the report to the Executive Board of Wageningen UR on 11 October 2001)

After Ms van Eck had given her address, the approximately 130 persons present had the opportunity to propose topics for workshops. Ten topics were put forward and those suggesting them gave a short explanation and reasons. The audience were then asked to decide which two of the ten topics should actually be made the topic of a workshop. Because of a malfunction in the voting equipment, it was not possible to produce a definite result but four topics were clearly favoured. Those proposing them all had the opportunity to implement their proposal. The FMD workshop was one of those proposals.

1.4.1 Overall methodology for the Foot and Mouth Disease workshop

The work of organising the Foot and Mouth Disease workshop began approximately one year after the outbreak. Heated discussions were still going on between many of the groups involved, but there was hardly any real dialogue. People keeping animals as a hobby accused the authorities of making a one-sided choice in favour of agricultural producers, while others pointed out that the psychosocial effects had been underestimated and that the aspect of financial settlement was no simple matter. In short, it did not seem a good idea to simply bring together the various different parties. A decision was therefore taken to start interviewing the main stakeholders on an individual basis. The interviews confirmed the enormous differences of opinion between the various parties involved, but they also made clear that they had not acted solely on the basis of their own interests during the epidemic. Each of them had basically acted in a manner which it perceived as being as far as possible in the public interest. It was therefore necessary to establish a methodology for study and analysis that would do justice to the complex nature of the crisis, that would allocate a role to each relevant party and opinion, and that could also be effective during the actual workshop. It was decided to create a problem tree (a core tool in the "Logical Framework Approach"; see www.snvworld.org/ruralDev/rural-development/oopp.htm) and to present it to two workshops, one for those directly involved and the other for researchers. The results are set out below.

The Foot and Mouth Disease workshop was one of the ways in which Wageningen UR responded to the Foot and Mouth Disease crisis of 2001. Another was the evaluation carried out by the Social Sciences Group, which reviewed the policy pursued during the outbreak. The study proposes multi-criteria analysis as a tool for use in future decision-making. A summary of this project is given in Chapter 3.

References

Bieleman, J., 1992 (In Dutch). Geschiedenis van de landbouw in Nederland 1500-1950. (Dutch agricultural history 1500-1950). Boom, Meppel, The Netherlands.

Cuijpers, M.P. and K.J. Osinga, 2004. What lessons have we learnt for the future? Position of the Dutch Farmers' Union LTO-Nederland regarding the future prevention and control of FMD. (www.lto.nl/dossiers/mkz/LTO-OIEcontribution.doc)

Frouws, J. and P. Leroy, 2003 (In Dutch). Boeren, burgers en buitenlui. Over nieuwe coalities in het landelijk gebied. (Farmers, citizens and villagers. About new co-allitions in rural areas). Tijdschrift voor sociaalwetenschappelijk onderzoek van de landbouw 18: 90-103.

LNV, 2003 (In Dutch). Debat Toekomst Intensieve Veehouderij (Debate future of Intensive Livestock Production). Ministry of Agriculture, Nature and Food quality, The Hague.

Maslov, A., 1954. Motivation and personality. Harper and Row, New York, USA.

Oosting, S.J. and I.J.M. de Boer, 2002. Sustainability of organic dairy farming in the Netherlands. In [I. Kyriazakis and G. Zervas, eds.] Organic meat and milk from ruminants, pp. 101-107. EAAP publication 106. Wageningen Academic Publishers, Wageningen, The Netherlands.

RLG/RDA, 2003 (In Dutch). Dierziektebeleid met draagvlak. Advies over de bestrijding van zeer besmettelijke dierziekten. Deel 1 en deel 2. (Infectious animal disease policy with public support). Raad voor het Landelijk Gebied, Amersfoort; Raad voor Dierenaangelegenheden, Den Haag.

Taskforce Waardevolle Landbouw, 2001 (In Dutch). Naar een waardeNvolle landbouw. (Towards a valuable agriculture). Wageningen UR, Wageningen, The Netherlands.

Van Broekhuizen, R., L. Klep, H. Oostindië and J.D. van der Ploeg, 1997 (In Dutch). Atlas van het vernieuwend platteland. Tweehonderd voorbeelden uit de praktijk. (Atlas of a transforming countryside). Misset uitgeverij bv. Doetinchem, The Netherlands, 240 p.

Van der Ploeg, J.D., A. Long and J. Banks, 2002 (Editors). Living countrysides. Elsevier bedrijfsinformatie bv. Doetinchem, The Netherlands, 230 p

Van der Weele, C.N., V. Beekman, M.M.M. Overbeek, S.L. Koole and C.W.M. Giesen, 2003 (In Dutch). WAVE (Waarden in vergelijking) (Values in comparison). Rapport 7.03.08. LEI (Institure of Agricultural economics), The Hague, The Netherlands.

Ziel, T. van der, 2003 (In Dutch). Verzet en verlangen. De constructie van nieuwe ruraliteiten rond de mkz-crisis en de trek naar het platteland. (Resistance and desire. The construction of new ruralities around the FMD-crisis and the migration to the countryside). PhD-thesis Wageningen University, Wageningen, The Netherlands.

Chapter 2.   The Foot and Mouth Disease workshop: from dialogue to research agenda

2.1 Summary and recommendations

2.1.1 Summary

According to the recommendations of the "Valuable Agriculture Taskforce", Wageningen University should focus more on the social role it can fulfil. The Wageningen Workshops in which scientists and stakeholders analyse problems and formulate solutions are part of this process. This is the report on the Foot and Mouth Disease (FMD) Workshop.

The FMD epidemic in 2001 became a crisis because insufficient attention was given to the social consequences of the epidemic. The crisis was analysed in the workshop, solutions were formulated and a research agenda for Wageningen University and Research Centre proposed. During this process, social actors have gained more insight and understanding of one another's points of view.

The FMD Workshop process comprised a literature analysis, orientation interviews alternated with two workshops – one for stakeholders and the other for scientists - reporting and reflection by the steering committee.

In the stakeholder workshop, a problem analysis was carried out and presented in the form of a problem tree, which shows the factors contributing to the outbreak of the epidemic that evolved into a crisis. Then the stakeholders presented solution pathways on which they considered research should focus. They also indicated criteria that the solution pathways must meet:
- Support by society
- Technically, process-wise and legally feasible
- Cost efficient (financially)
- Integrated weighing of the consequences of the solution for different sectors such as nature, recreation, small and medium-sized businesses, agricultural enterprises
- Contribution to sustainability
- Ethically sound
- Cause minimum emotional stress.

With these criteria in mind, the stakeholders formulated five solution pathways, in the following order of priority:

- A completely new scenario (strategy, procedures and protocol) for dealing with a FMD outbreak (16)
- Renewed discussion in the EU framework (12)
- A pilot scenario to be tested in one area (7)
- Examination of the production chain for processing products from vaccinated animals (7)
- Conversion of the deduction system to a penalty system with regard to financial support from government, based on qualifications of farmers. (4)

The numbers in brackets indicate the number of votes each of these solution pathways received.

The scientists' workshop resulted in various additions to the problem tree and identified the current focus areas of research on FMD. This clearly indicated those problems not receiving sufficient attention. These 'weak' spots were identified as:

- International framework for policy innovation
- Future role of the livestock industry in rural development in relation to FMD control
- Attention to the production chain and consumers and the effect of both on policy
- Social cost/benefit analysis of the FMD crisis
- Interaction between scientific disciplines particularly with respect to communication
- Process of management decision models, conduct and style of action during an outbreak and in the preparation for control measures.

The scientists proposed a research agenda for the solution pathways put forward by stakeholders and for one of the "weak" spots identified.

The solution pathways proposed by the researchers and discussed by the sounding board group with representatives of business and social organisations were formulated into three recommendations. These recommendations were prepared on the basis of knowledge and recent reports on developments in the livestock industry and on FMD outbreaks in the Netherlands and the United Kingdom.

2.1.2 Recommendations
1. Wageningen University and Research Centre takes up the challenge of formulating a research agenda to support knowledge requirements for policy preparation and political decision-making by farmers' organisations, social organisations and governments.
2. The research agenda recognises the importance of existing research, identifies new areas for research and takes up the challenge of an interdisciplinary approach.
   The agenda comprises:
   - Organise or be actively involved in a bottom-up FMD analysis of the context in order to incorporate social changes in a current scenario and to up date this scenario every three years. The opportunity to re-consider FMD free-status in the future must be retained. Research the quality of such interactive processes.
   - Be actively involved in or organise a pilot study or exercise at local level with the relevant government bodies and social actors, using the lessons learned from the recent crisis. For this purpose, an analysis of the local context is essential in order to be able to call upon local knowledge and expertise. This local knowledge must be tested locally in an exercise to ensure that a national scenario including its procedures and protocols fit the local situation.
   - Research and evaluate different scenarios for emergency vaccination in managing future outbreaks. Give attention to the veterinary, economic, trade politics and psychosocial consequences of an outbreak. Research the commercial feasibility of local processing of products from FMD-vaccinated animals. Also, research the potential for local standstills and quarantine without the rest of the country losing its FMD-free status with regard to exports.

- Carry out epidemiological studies on the transmission and spread of FMD in relation to infrastructure and transport. Stimulate further development of a marker vaccine and rapid diagnostic techniques.
- Research the potential for improving the effectiveness of Dutch and European management and policy. Research the potential for different ways of executing FMD policy with regard to different types of enterprises, regions and countries within the European Union. Knowledge is needed about the rural economy of different regions and countries in the European Union, to which control strategies must be adjusted. Research the management process in different EU countries.
- Research communication before and during an outbreak and a crisis. The responsibility does not lie entirely with the Ministry of Agriculture, Nature Conservation and Food Quality but also with other ministries, social actors and scientists. Communication from the different government bodies has to be attuned. A rapid analysis of the surroundings with a few people who know the area, at the beginning of a pilot study, can help determine and invite the relevant partners at each management level. The role that various communications means (medium, place and time) can play when there is a crisis and no crisis should be determined.
- Make a proposal for a EU-funded research project in which the Member States adopt the approach of the FMD Workshop. Use the results to develop a future EU strategy.
3. Centre for Interactive FMD research to carry out an extensive research agenda should be set up.
- The centre must operate independently of other science units and be responsive to the WUR Board of Management. This expert centre must organise innovative, interdisciplinary research on FMD and possibly other infectious animal diseases.
- Continuity must be ensured in the preparation and implementation of research together with the stakeholders. Contact with this group is essential to ensure the social relevance of research.
- The researchers involved in this centre must have, in addition to the subject matter expertise, an active interest in and skills for interactive research, and communication skills for cooperation with other disciplines and with stakeholders.
- Contacts must be developed with research centres and other universities. The presence of Wageningen and the effectiveness of an integrated research programme must be strengthened.
- MSc students should participate interactively in carrying out the research.

2.1.3 Conclusions

The workshop aimed to stimulate dialogue between stakeholders and between stakeholders and scientists in order to achieve a socially responsible research agenda for the Wageningen University. This interaction has successfully led to the formulation of a research agenda. Also, a start has been made on dialogue between the various stakeholders during the workshops but there was only limited dialogue between scientists and stakeholders. The workshop is a first step in stimulating this dialogue and initiating further action from the dialogue. The Taskforce Valuable Agriculture concluded that research with an interactive, interdisciplinary and innovative approach must be stimulated within the Wageningen University and Research Centre. The FMD Workshop has stimulated a new approach to formulating a socially responsible research agenda.

The research agenda must lead to research that has added value for future policy making. Interactive research will have more meaning to policy makers in terms of integration and harmonisation, saving time, and methodology combining disciplines. In the FMD Workshop, research policy was developed in consultation with stakeholders, scientists and policy makers.

2.2 Introduction

2.2.1 Foot and Mouth Disease epidemic

There have been several outbreaks of Foot and Mouth Disease in the Netherlands in the past. The last large-scale outbreak of the disease was in 1966-67. During and after that epidemic, vaccinations against FMD were carried out annually. After that time, FMD have occurred only sporadically and after 1984, there were no further outbreaks of FMD in the Netherlands (Bieleman, 2001). In 1992, it was decided to stop vaccinating cattle against FMD in the EU. This decision was supported in the Netherlands based on a study that weighed the economic loss against the cost of FMD vaccination (Berentsen *et al.* 1990).

Following the notification of FMD in the United Kingdom on 19 February 2001, the Netherlands was infected with the virus on 21 March 2001.This occurred because imported calves came into contact with infected sheep at a rest and watering place in Mayenne, France. The farming enterprises to which these calves were delivered were cleared as a preventive measure but still FMD broke out in Fortmond in Olst Municipality (Overijssel). In the period between 21 March and 22 April, there were further outbreaks in the Veluwe near Oene, Oosterwolde, and Kootwijkerbroek; in Olst and Wijhe; and in Friesland, Ee and Anjum. A total of 26 farms were infected and about 3,000 farms including 1800 hobby farms were cleared as a preventive measure. The clearance was completed on 23 May with 265,000 animals slaughtered, including 93,000 cattle, 118,000 pigs, 35,000 sheep, 8,000 goats and 11,000 other animals. Of these animals, 65,000 were slaughtered and 200,000 were slaughtered after emergency vaccination. On 26 June, two months after the last animals were slaughtered, the last area, the Veluwe, was declared FMD-free and restocking could begin there (Siemens, 2001).

The economic loss due to the FMD crisis has been estimated by the Agriculture Economics Institute (de Bont & Wisman, 2002) at approximately Euro 800 million. This amount only includes the cost to livestock farms in the form of lost income and mandatory contributions to animal health funds, national governments and the EU. The amount does not include damage and loss of tourism and loss to other sectors and organisations. When this is included, most likely, the total loss would be well in excess of one thousand million euros.

2.2.2 Wageningen Workshops

The Wageningen Workshops are the initiative of the Taskforce of the Wageningen University and Research Centre (WUR). The Taskforce was a group of scientists who advised the Board of Management on future research directions and methods. They proposed analysing problems and consider solution pathways together with stakeholders and scientists. One of these knowledge workshops is the FMD workshop, which was proposed by Prof. Dr A. J. van der Zijpp. This workshop was to be a forum where stakeholders and scientists could enter into dialogue and where both parties could determine the direction of the research. The workshop aimed to produce an integrated research agenda for WUR and partners and built on the wishes of the stakeholders. WUR wants these workshops to intensify interaction between scientists and society. The research agenda can also be determined from the outcome of the workshops.

2.3 Purpose of the workshop

2.3.1 Objective

The purpose of the workshop was to formulate a research agenda on FMD for Wageningen University and Research Centre. By involving stakeholders in setting the research agenda, the university is increasing its social involvement and giving practical meaning to its research.

The workshop objective was to analyse all aspects of the complex problem of FMD jointly with stakeholders and to learn from their experience. This analysis can support strategic decision making on future research, policy and politics.

Time schedule

The activities of the FMD Workshop are set out in Figure 1.

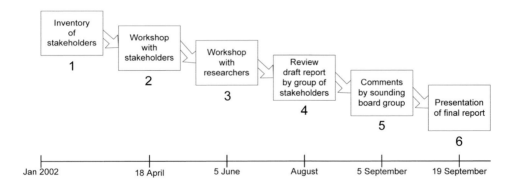

Figure 1. Time schedule for FMD Workshop.

1. Inventory by stakeholders   January 2002
   In order to structure the FMD Workshop, an inventory was made of the issues based on a literature analysis (see reference list) and interviews with stakeholders (Appendix 4). These two sources were used to make the initial problem tree and in organising two workshops.

2. Stakeholders Workshop   18 April 2002
   The first workshop with stakeholders (see Appendix 4) took place on 18 April 2002. The workshop aimed to discuss the problem from different perspectives and to exchange experience in a joint analysis of the FMD crisis. The analysis is presented in the problem tree, based on the ways of preventing a new outbreak and alternative approaches to dealing with a new outbreak. This approach examined the technical and financial consequences of the crisis and also the underlying cost to man and animal.

21

Also, criteria for assessing possible solutions were considered and used by the participants to set priorities for solution pathways.

3. Researchers Workshop   5 June 2002
   The second workshop with scientists (Appendix 4) took place on 5 June 2002. The workshop made an inventory of current research on FMD in the Netherlands, and identified areas where little or no research has been carried out ('weak spots'). Furthermore, three solution pathways identified in the first workshop and one 'weak spot' in the research issues were discussed and also the research expertise required.

4. Review of draft report by stakeholder representatives   August 2002
   In August 2002, the draft report was presented to a group of stakeholders who had attended the first workshop. The report was reviewed to ensure that it adequately reflected the thoughts and ideas expressed in the first workshop.

5. Comments by sounding board group   5 September 2002
   The draft report was also presented to the sounding board group (Appendix 4) on 5 September 2002 for detailed comments on the recommendations set out in the report.

6. Presentation final report   19 September 2002
   The final report was presented to the WUR Board of Management on 19 September 2002.

2.3.2 Approach

The FMD Workshop set up a steering group to guide the process from the perspective of different research departments. Scientists from departments and institutes in Wageningen University and Research Centre were invited to join the steering group. The steering group met for the first time after the first workshop with stakeholders. The members are Ms. I. de Boer and Prof. A. van der Zijpp representing Animal Production Systems Group (APS); Ms M. Mourits, Business Economics Group (FMA); Ms N. Aarts, Communication and Innovation Studies (C&I); Mr. R. Schrijver, ID Lelystad and Mr B. ten Hoope, organisation consultant from the university Management Centre. After the second workshop with scientists, the steering committee was extended to include Mr T. Vogelzang, Agricultural Economics Institute (LEI), Mr H. Kieft (ETC, facilitator), Mr T. van der Ziel (rural sociologist) and Mr S. Oosting (APS). In addition to guiding the process, the steering committee advised on further developments and planning of the FMD Workshop, and the final report.

The workshops were facilitated by Mr H. Kieft from ETC, Leusden. The sub-working groups during the workshops were led by members of the steering committee. The reports during the workshop were prepared by students and graduates of the Wageningen University.

The Wageningen Workshops aim to facilitate dialogue between society and scientists. It was, however, decided to carry out separate meetings for stakeholders and scientists. This gave the stakeholders the opportunity to analyse the FMD outbreak and crisis from different perspectives. Separate workshops were also organised to prevent the dialogue between those socially involved becoming dominated by the discussions between scientists in different disciplines. Holding the stakeholders workshop ensured that the scientists started with the solutions, priorities and criteria of the stakeholders. Because it was anticipated that the stakeholders would come up with interdisciplinary solutions, the scientists had to link up with these issues and propose innovative, interactive and interdisciplinary research topics.

Cooperation within the Wageningen UR

At the time of organising the FMD Workshop, the Department of Social Sciences, Wageningen University, was carrying out a research project entitled "Prevention and Combating Foot and Mouth Disease". Other institutes involved in this project are the Agriculture Economics Institute (LEI), Business Economics Group (FMA) and ID Lelystad. As LEI had carried out a stakeholder analysis in this research project, it was decided to work together and to share the interviews in the stakeholder analysis. LEI was, thus, closely involved in the activities of the workshop. A summary of this research project is given in Chapter 3.

2.4 Problem analysis

2.4.1 Method

The problem analysis is based on the problem tree method (www.web.mit.edu) that sets out diagrammatically the cause and effect relationship of sub-problems, which together make up the main problem. In this approach, the main problem (in this case, the FMD outbreak and crisis) is defined and the causes and effects of the problem formulated. The problem is always presented as a noun and verb, and linking the sub-problems to their causes and effects. This places the problem in a wide context, enabling relationships to be established between the various aspects of the problem. The problem tree is then reformulated into an objective tree by expressing problems as solutions. This process reveals which problems must be solved and the effect that will have on other sub-problems. In addition, new opportunities and solutions can be sought that do not flow directly from converting the problem tree to an objective tree.

2.4.2 Result

The final FMD problem tree was developed in several steps (see Appendix 1). The initial problem tree was prepared from a literature review and interviews with stakeholders and then verified and extended in the stakeholders workshop. The participants worked in subgroups to extend and add to the branches and roots of the problem tree. The extended tree was used in the scientists workshop to show the current research focus and to identify gaps in the research. Subsequently, the problem tree has changed very little, although a few additions were made later. The problem tree should not be considered as complete and its logic can always be improved but it does represent the perceptions of the FMD Workshop participants.

The simplified problem tree (Figure 2) shows there are two main problems: the outbreak of the FMD epidemic and as a consequence, the FMD crisis. The FMD epidemic comprised four key factors that can be traced to a set of sub-factors (Appendix 1), namely (1) International trade policy; (2) International and Dutch decision-making; (3) FMD is endemic around the world; and (4) Inadequate control and prevention. Factors 1 and 2 have led to the non-vaccination policy, while factors 3 and 4 to the permanent risk of infection. This combination of factors led inescapably to the outbreak of the FMD epidemic.

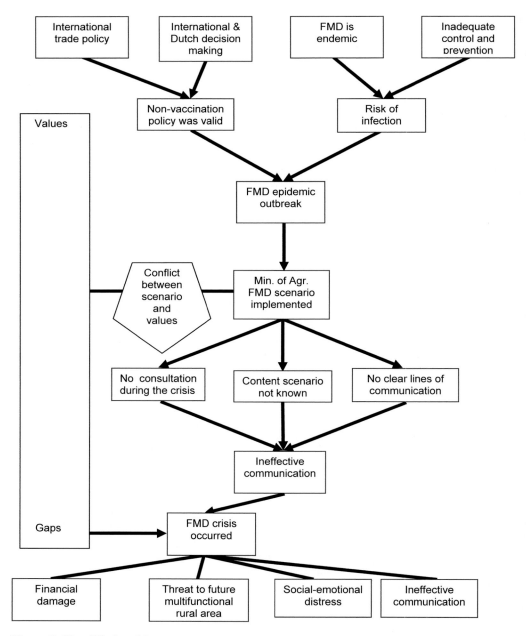

Figure 2. Simplified problem tree.

The problem tree shows why the FMD epidemic developed into a crisis. The shift in the values of society and agriculture, only became clear from the reactions to the outbreak and control measures taken. The scenario of the Ministry of Agriculture, Nature Conservation and Food Quality was mainly based on veterinary-economic considerations and the societal values at the beginning of the 1990s. Awareness of a shift in values has clearly increased. In addition, communication with farmers, hobby animal owners and social organisations during the crisis was not considered adequate. The problem tree also shows the consequences of the crisis, particularly the fact that there were financial consequences not only for livestock producers but also for other parties such as tourism, and medium and small companies in rural areas. It also shows that the future for multifunctional rural areas was at risk (with many hobby animal owners), that communication was not effective, and that those directly involved suffered social and emotional distress. The causes and the consequences (see Social Emotional Stress) of the social and emotional distress are presented in the extended problem tree (see Appendix 1).

2.4.3 Results

Stakeholders workshop

The participants were very appreciative of the opportunities to meet and dialogue with others involved in the FMD crisis. Various groups communicated with one another. The participants considered the workshop to be useful and that it should have taken place much earlier. However, there was a notable lack of knowledge about FMD and how the virus is transmitted.

Although representatives of industry were invited to attend this workshop, very few did. This was considered a loss by the other participants. As a result, discussions tended to be rather one-sided on topics, such as international trade and consumption of products from FMD-vaccinated animals and the point of view of industry.

Outcome of stakeholders workshop

The workshop aimed to make a joint analysis of the complex problem and to indicate the solution pathways to which research and policy should be directed. The results of the problem analysis are presented in Appendix 1.

The participants discussed the solution pathways in four working groups. The discussion topics had been defined in advance of the workshop: communication, social-emotional distress, multifunctional rural areas and international policy. The solution pathways proposed by the working groups overlapped one another. As a result, the solution directions were summarised and adjusted in a plenary session. These solution pathways were presented in the language and terminology of the stakeholders.

The participants formulated their priorities for the proposed solution pathways on the basis of the agreed criteria. These criteria were initially defined on the basis of literature reviews and interviews with stakeholders and later adjusted by the workshop participants.

Criteria to be met by the proposed solution pathways
Supported by society
Technically, process-wise and legally feasible
Cost efficient (financially)
Consequences for sectors such as nature, recreation, small and medium-sized businesses agricultural businesses
Contribution to sustainability
Ethically sound
Cause minimum of emotional distress.

The following observations were made during the workshop:
- Three groups played a role in communication about ministerial procedures in the scenario regarding the FMD crisis, namely the Ministry of Agriculture, Nature Conservation and Food Quality, producers of milk, meat and eggs, and stakeholders. A distinction can be made between those involved with animals, such as dairy farmers and hobby farmers, and those involved without animals such as medium and small businesses and the catering industry. The information provided was factual, with insufficient insight, and there was no opportunity for feedback to the Ministry.
- The regional responsibility taken by the municipalities and local organisations (role of the churches, and social organisations) during the crisis was greatly appreciated and acknowledged.
- The farmers and other stakeholders did not know the content of the procedures in the FMD scenario and the reasons behind them. As a result, the purpose of the measures was not understood. The procedures were based on veterinary expertise in FMD, and an extra communication input was required to those involved and to farmers, who in most cases were less familiar with them. In addition, the procedures were based on values patterns of ten years ago, which appear to have shifted since then. This contributed to the lack of understanding of the measures taken.

The solution pathways chosen by stakeholder workshop participants were:
1. A completely new scenario to deal with a FMD outbreak (16)
2. Open discussion about non-vaccination in the EU framework (12)
3. A pilot scenario developed and tested in one area with stakeholders (7)
4. Re-evaluation of the production and sale of FMD-vaccinated animals (7)
5. Conversion from a deduction system to a penalty system (4)
6. Strengthening border controls and fast closures in the event of suspected cases of MDF.

The number of votes for each solution pathway is given in brackets. The first four solution pathways were used in the stakeholder workshop as input for the research agenda.

These solution pathways were proposed by the stakeholder workshop and thus were determined on the basis of the knowledge of those present. As no industry representatives attended the workshop, the solution pathways were not discussed from that perspective. Solution pathway 6 - stricter border controls - did not receive any votes after the discussion because the participants from the Ministry and the Commodity Board PVE explained that for the Netherlands, this would mean stopping exports on average 20 times a year because of suspected cases. Possibly, an explanation from the industry about solution pathway 4 - selling products of FMD-vaccinated animals - could have further influenced this solution pathway.

Experience in the scientists workshop

The workshop revealed an uneven distribution of scientific expertise throughout the problem tree (Figure 3). The emphasis was on research about and around the ministerial procedures, particularly veterinary and economic aspects. The circles indicate concentration of research areas; the size of the circle is not relevant. Little research has been done on the social consequences of the FMD crisis. There appears little or no interaction between the various disciplines. During the workshop, scientists appeared to have difficulties in adopting an integrated approach to the issues and often continued to focus on their own discipline.

Further, Wageningen UR does not have all the expertise needed in tackling the complex problem of FMD. Several disciplines required in a integrated approach were not represented such as business administration, communication and politics.

It was also clear that both stakeholders and scientists needed to have empathy with the problem in order to understand its complexity and to jointly investigate the issues. Attention must be given to this in the training and development of scientists.

Outcomes of the scientists workshop

The scientists workshop aimed to extend the problem analysis by making an inventory of current research on FMD and to identify subjects not being investigated ('weak spots'). In addition, the participants defined the research issues and expertise needed for the three solution pathways identified in the stakeholders workshop and the 'weak' spot's identified after the current research was placed in the problem tree.

Areas related to FMD in which little or no research is being carried out:
A. International framework for policy innovation
B. Future role of livestock industry and rural development in relation to FMD control
C. Production chain and consumers and their effect on policy
D. Social cost/benefit analysis of the FMD crisis
E. Interaction between scientific disciplines specially in communication
F. Prevention rather cure management of an epidemic
G. Administration and management decision models, conduct and behaviour before, during and after an outbreak.

The participants proposed research issues for the four solution pathways identified as priorities by the stakeholders workshop. Solution pathway 1 - New procedures must be established to management a FMD epidemic - was combined with Solution pathway 3 - A pilot scenario be tested with stakeholders in one area. Solution pathway 2 – Non-vaccination policy to be discussed in a EU framework - and solution pathway 4 - Re-evaluation of production and sale of products from FMD-vaccinated animals - was a separate topic of discussion. The scientists added a fourth discussion group the 'weak spot' G - Administration and management decisions model, conduct and behaviour before, during and after an outbreak.

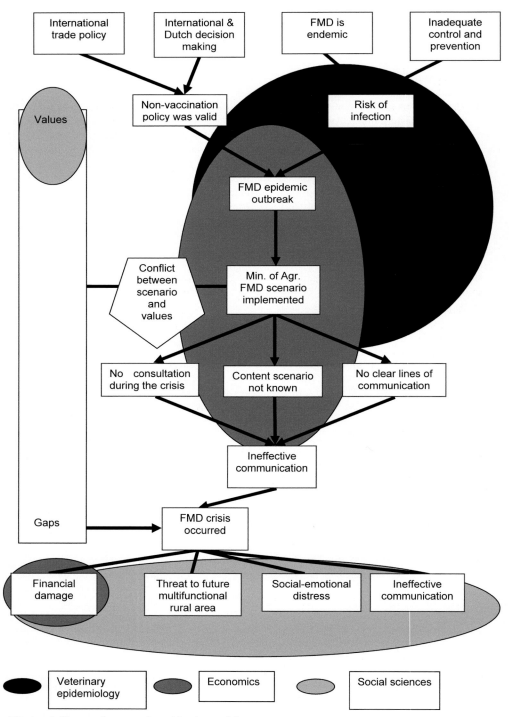

Figure 3. Research areas placed in the problem tree.

The research issues were developed in four working groups and presented in a plenary session. Beyond this, there was no further discussion about the research issues proposed by the other participants. The research issues presented in the workshop are as follows:

1. 'New procedure to manage a FMD epidemic combined with a pilot scenario tested with stakeholders in one area'.
   - Carry out an extensive analysis of the context in which all relevant social organisations are actively involved. On the basis of this analysis, a new procedure should be developed that is understood and supported by as many parties is possible. Use the experience from the pilot study to adjust the analysis of the context. Process-directed research can improve the quality of cooperation between government and the local residents.
   - Starting from the current non-vaccination policy, various scenarios for emergency vaccination should be set up and evaluated in cooperation with the various stakeholders.
   - Communication must be improved between the Ministry of Agriculture, Nature Conservation and Food Quality, local governments, social organisations and agriculture organisations before and during an FMD outbreak.

2. 'Non-vaccination policy to be discussed in the EU framework'.
   - Seek to link the European FMD policy with the farming styles (for example, multifunctional or specialised) in different European countries. FMD policy in the European Union must allow for different farming styles and regional differences. Research, therefore, farming styles in relation to controlling FMD outbreaks and carry out risk analyses.
   - Develop alternatives for controlling FMD with regard to political decision-making in Brussels. This will require insight into complex decision-making processes.
   - Analyse the cultural basis for political decision-making in the Member States including the aspirant Member States. The emphasis must be on controlling infectious animal diseases in general.
   - The European Union must contribute to the development of animal production outside the EU in order to prevent or limit the movement of animal diseases such as FMD.

3. 'Re-evaluate production and sale of products (dairy, meat) from FMD-vaccinated animals'.
   - Research the commercial feasibility of processing products from vaccinated animals.
   - Develop a guarantee system for the monitoring and distribution of products from FMD-vaccinated animals in the Netherlands without endangering the country's export position.
   - Carry out a study on the development of a marker vaccine that will enable a distinction to be made between field infection and vaccination. Research is also necessary on rapid diagnosis for a marker vaccine and the field virus. The risk of FMD-carrying animals must be determined.
   - Study the chances and limitations of corporate governance for production focuses for the internal market. Define the risks of globalisation and international competition (open borders).

4. 'Administrative and management decisions model, behaviour before, during and after an FMD outbreak.'
   - Obtain more insight into 'bottom-up' initiatives in order to prevent FMD control being largely approached objectively (technically, measurably, and centrally controlled) with insufficient attention to the human aspects at various policy levels.
   - When local stakeholders become more central, the decision-making process should become more interactive. Communication is important but also the context namely the social debate about animals and animal disease control. This discussion will determine the extent of support for new FMD regulations. Research must indicate whether and how this support can be attained.
   - Achieving consensus or consent (not everyone will be happy with all aspects but will have to accept that) is one of the greatest challenges for the new policy. Research is needed on how society is re-oriented and organised for people individually and collectively, and how management decision models can be adjusted.

2.5 Towards an integrated research agenda

2.5.1 Introduction

The threat of FMD is always present because the disease is endemic worldwide and because of the increasing movement of people, animals and animal production. Therefore, thorough account must be taken of the repeated threats and outbreaks of FMD in the future.

The FMD outbreak in 2001 was approached as a technical, veterinary and epidemiological matter and the epidemic was brought under control in a relatively short period and with praise from the Office International des Epizooties (OIE). Socially, however, this approach caused a lot of distress. The approach led to considerable resistance both within and outside the agricultural sector and is thus not justified in the future. The main reason for the resistance was that opinions about animal welfare have changed considerably since 1990 when the ministerial procedures were formulated largely based on veterinary and economic considerations. This value conflict was experienced by both livestock producers and livestock clearing teams as well as by hobby animal owners and local inhabitants. The social debate about killing healthy animals and killing healthy FMD-vaccinated animals has raised many moral and ethic issues. In addition, poor communication from the Ministry of Agriculture, Nature Conservation and Food Quality to the various groups of stakeholders played a significant part in resistance to implementation of the ministerial procedures (COT, 2002).

The crisis after the FMD outbreak in 2001 should also be seen in the light of the changing role of rural areas in the Netherlands. Rural areas are not only for agriculture and animal production, but are increasingly important in nature development, recreation and as residential areas for the urban population. The countryside is also important for small and medium enterprises. There are changes in animal production, It is no longer the only source of income for farming enterprises, which have expanded to tourism, human care, nature management, processing of products (Van der Ploeg *et al.*, 2002), involving people outside animal production in animal production. As a result, not only farmers but also the rest of the population have an interest in the development of the countryside. There is a large group of hobby animal owners in the countryside with their own specific interests. This was obvious in the FMD crisis, which also affected the tourism sector and medium and small business enterprises (The Economist, 2002).

Research has focused in the first instance on evaluating the 2001 crisis in the Netherlands (B&A Groep, 2002; COT, 2002; Dierenbescherming, 2002; KNAW, 2002; LTO, 2001; Van Haaften en Kersten, 2002) and in the United Kingdom (The Royal Society, 2002; Anderson, 2002). From the mistakes made, lessons for the future must be extracted. The FMD crisis has also brought the future of animal production under discussion (Denkgroep Wijffels, 2001; Taskforce Waardevolle landbouw, 2001; Curry, 2002). In addition to research on FMD in the strictest sense of the word, a broad analysis is needed of the development of animal production in the Netherlands. This must be the background to integrated research on FMD. In the future, research must focus on:

1. Programme of analysis of the context specific to the FMD area, including social aspects and functional networks;
2. Epidemiological research on FMD;
3. Pilot exercises and studies at local level with active involvement of stakeholders (for a North American example, see www.inspection.gc.ca).
4. Prevention and control of future outbreaks and the link with ongoing changes in societal values.
5. The effectiveness of Dutch and European administration and policy
6. Communication before and during an outbreak so that the mistakes of the past are not repeated in a future FMD outbreak.

These findings together with the outcome of the two workshops have resulted in a proposed research agenda for WUR. The research agenda must also address the policy, for example, of LTO (Farmers Union), Ministry of Agriculture, Nature Conservation and Food Quality, Animal Protection Society, EU and other organisations.

2.5.2 Research agenda

1. FMD analysis of the context

The knowledge and views of animal producers and the countryside are constantly changing and the FMD crisis has reinforced this. Animal production and the countryside of the future in the Netherlands have been placed in a new light (Denkgroep Wijffels, 2001) and in the United Kingdom (Curry, 2002). The relationship between city and country continues to determine developments and what interventions are acceptable. During the FMD crisis, there was wide spread social debate about the slaughter of FMD-vaccinated animals. This covered public health (food safety) and economic issues (need to slaughter for the safety of the market) as well as other functions that animals such as food production, and the role of animals in sport, nature management, recreation, human care and hobbies. In those areas where the animals were slaughtered, attention needs to be given to the social-emotional consequences of such a crisis not only for farmers but for society as a whole.

In order to gain more and continuing insight into what is happening in society, we propose to carry out an analysis of the context and specifically on the consequences of changes with respect to FMD policy, and to update this analysis every three years. This analysis must be carried out with people in the area, using a 'bottom-up' participative method. This will show continuing changes, such as in norms and values so that these changes and possibly also trends can be included when FMD procedures are updated. This analysis can support policy development and so ensure that the FMD procedures meet the needs of the time when they are applied.

By regularly updating the context analysis, insight can also be gained in the process of psychological ´recovery´ and the relationship between policy measures and government initiatives and social actors.

In addition, maintaining FMD-free status should be re-considered at regular intervals. Insights into aspects, such as risks (for example, transport of people and animals, and developments in hobby animal farms), better vaccines, and international trade policy agreements, will all have an effect on maintaining this disease-free status.

2. Pilot exercise and study at local level

Before a pilot exercise is carried out at local level, a wide social and scientific evaluation of the 2001 crisis (B&A Groep, 2002; COT, 2002; LTO, 2001) needs to be carried out. Much can be learned from how the crisis developed in 2001 and this information used in setting up a pilot study. Local support was greatly appreciated in the 2001 crisis and possibly more effective use can be made of it. Further, a local analysis of the context needs to be carried out to determine the stakeholders in various areas. Their knowledge and expertise should be sought and used in the pilot study. But it is not just their knowledge. The input of some local persons and organisations in formulating region-specific procedures, in the pilot exercise and possibly in its implementation, can considerably improve the quality and acceptance of measures to be taken. This knowledge and effort should be tested in an exercise and could result adjustments to the general scenario of procedures and protocols to meet the needs of a specific local situation.

The Royal Society (2002) has sketched such an approach with four practical steps: horizon scanning, contingency planning, rehearsal and learning. Following these steps systematically will reveal any gaps in knowledge, local networks, means, and capacity so that improvements can be made to the ministerial procedures.

3. Research on controlling a future FMD outbreak

Quantitative models need to be developed in order to develop various scenarios for emergency FMD vaccination taking into account the veterinary, economic and psychosocial consequences of an outbreak. The scenarios should also include the consequences of emergency vaccination with regard to exports and the period before the 'all-clear' can be given for an area by the OIE/WTO (3-6 months).

Economic and decision-making models should be developed in order to weigh social, ethic and psychological factors against business and financial factors (KNAW, 2002) and used in selecting the most appropriate scenario at the time of an outbreak.

In addition, the commercial feasibility of processing products from FMD-vaccinated animals should be determined. An 'early warning system' (The Royal Society, 2002) must be developed for better and faster tracking of the transport of animals and animal products to enable earlier invention when necessary. This must be set up in an international context (EU and also beyond) because of cross border transport.

The precautionary principle in the case of suspected FMD must be applied so that an area can be brought to a 'standstill' or quarantined earlier and faster without hampering exports. Here also economic and social considerations are important.

4. Epidemiological FMD research

More fundamental research needs to be done on the prevention, spread and transmission of the virus under various conditions and on all serotypes. The further development of a marker vaccine is important, and for the long-term, a vaccine would have to be developed that offers protection against all serotypes and gives life-long immunity. Research on a rapid diagnostic technique is vital. This could link, for example, the knowledge of PCR field diagnosis for all serotypes with GIS data, such as is now being developed in the United States. These diagnostic techniques are now being tested in Uruguay (P. Cunningham, pers. comm.).

More information is required about the prevention and spread of the disease in the field. This would lead to measures such as improved hygiene to prevent the spread of animal diseases being better understood and accepted. Farmers can take measures for their own farms and set up their own procedures. Special attention should also be given to the large group of hobby animal owners in the countryside.

An epidemiological evaluation of the 2001 outbreak will enable reflection on control measures taken and to be taken.

5. Research on improving the effectiveness of Dutch and European management and policy

Regulations for the control of animal diseases are largely determined by European legislation. The Dutch interpretation of these regulations is under discussion and there are variations in interpretation between the Member States. It must be clear what room there is for national and regional interpretation.

Scenarios for different countries within the EU must be developed in which emergency vaccination can be carried out, and whereby the emergency vaccinated area is closed, without hindering national exports within and outside Europe.

Knowledge is required about rural economics in the various EU countries, not only about agriculture and livestock production but also other activities such as tourism and medium and small businesses. Control strategies will depend on the composition of the rural economy in the different countries and regions. Knowledge about the different livestock sectors in a region is important for understanding the consequences of a possible outbreak.

The Agriculture Economics Institute (LEI) is carrying out research on the administrative and social decision-making processes in the various EU Member States.

6. Research on communication before and during an outbreak

During the 2001 crisis, it was clear that the FMD outbreak was no longer the sole responsibility of the Ministry of Agriculture, Nature Management and Food quality. Other ministries were involved including Health, Welfare and Sport, Economic Affairs, Transport, Public Works and Water Management, Home Affairs, Justice and Defence. It is therefore essential that policy and communication is coordinated between the different ministries. A rehearsal and pilot study in one region can determine which individuals, organisations, and government bodies should be involved if there is an FMD outbreak and what is expected of them. This should improve communication between these players at national, provincial and municipal level.

Research is needed on the role different communication means can play during a FMD crisis and also when there is no crises. On one side, this concerns the role of the mass media, and more small-scale, interpersonal forms of communication on the other. It is most important to determine what is the right method, at the right place, at the right moment. For example, experts on the social map of the region should be identified.

Attention needs to be given to the type of communication, where it should be used and with what media, and what personal qualities are required. Consideration has been given to the role of different types of media (LTO, 2001) in the 2001 outbreak and this should be studied in more detail.

We stress here that communications must be considered in the specific context with due attention to the activity and creativity of the party or parties to be reached. Communication can no longer been seen as the transmission of factual information without sufficient attention to the receiver's interpretation. We must be aware that interpretation also depends on what has happened earlier and what (in the eye of those involved) will happen in the future. In addition, communication research account must take account of the fact that not only what the sender consciously produces but what he unconsciously delivers also carries a message. Both non-verbal signals as well as actions can be interpreted differently to that intended.

7. Knowledge transfer to other EU Member States

Knowledge and experience in the Netherlands on the control of the 2001 FMD crisis needs to be transferred to other EU Member States that did not experience the FMD crisis, as well as to the United Kingdom. The most appropriate approach would be to formulate a research proposal based on the FMD Workshop for EU funding. The results can be used in future EU procedures and to give an indicate the degree to which independent decision-making is possible in each region.

2.6 Towards an integrated research group

Research on the complex problems surrounding FMD should meet three criteria - innovative, interdisciplinary and interactive. To achieve this goal, an integrated research agenda needs to be developed which integrates research activities on FMD in Wageningen University and Research Centre (WUR) and elsewhere. In identifying the 'weak' spots, it has become clear that WUR does not have all the required disciplines in house (for example, politics and business management). A Centre for Interactive FMD Research could grow to become an interactive research centre for infectious animal diseases. This would enable an integrated approach to research on swine fever, BSE and other infectious diseases.

The Centre for Interactive FMD Research should be directly responsible to the WUR Board of Management and thus not depend on input from other knowledge units. A structure should thus be provided within the WUR to enable the centre to function effectively. Applications for finance could be made to the Ministry of Agriculture, Nature Conservation and Food Quality-DWK, EU, Innovative Network Green Space and Agrocluster and other organisations. Funds for integrated research agenda should be administered by the centre. Part of the funds can be used for preparing tenders, which include knowledge and initiatives from other institutes inside and outside the Netherlands.

The Centre for Interactive FMD Research should comprise scientists from the WUR and elsewhere. The members of the research groups as well as knowledge and expertise must meet a number of complementary conditions.

The members must be able to communicate with researchers from other disciplines and estimate the added value of these disciplines because the interdisciplinary nature of the research must deliver added value.

Ongoing communication between scientists and stakeholders is essential to the success of an integrated research programme, and thus expertise in participatory techniques needs to be built up. The language and objectives of stakeholders should be translated into understandable topics for the scientists.

It is important that scientists in the centre develop good social skills because effective interaction with the work field and stakeholders will deliver added value.
Special attention must be given to the issues of policy making organisations and to the needs of political decision-making.

As a starting point for the Centre for Interactive FMD Research, the steering group of the FMD Workshop could work out a proposal for the WUR Board of Management.

2.7 Evaluation of the FMD Workshop

2.7.1 Introduction

As well as the experience with the 2001 FMD crisis, there is also experience with Wageningen Knowledge Workshops. The evaluation focused on three aspects:
conclusions regarding the research agenda; dialogue in the FMD Workshop; and the process of the FMD Workshop. The process and the experience of the FMD Workshop are set out in the process report. The main conclusions are presented here.

2.7.2 Conclusion about the research agenda

Objective

The FMD Workshop aimed to stimulate dialogue among stakeholders and between stakeholders and scientists to reach a socially responsible research agenda for the WUR. This has led to formulating an interactive research agenda.

Links with ongoing research

The outcomes of the FMD Workshop must be linked to current research plans. The original concept of an integrated approach to achieve a socially accepted solution was re-enforced by the workshop. Research now being carried out in this area at WUR, however, lacks an integrated approach. Efforts need to be made to reduce the risk that scientists merely select their research topics from the integrated research agenda proposed. Researchers need to actively contribute to an integrated approach and thus ensure the social and societal significance of the work.

2.7.3 Conclusions about the dialogue

Objective

Dialogue between stakeholders started during the workshops, but dialogue was limited between stakeholders and scientists. The FMD Workshop did not have the means and the time to actively initiate and facilitate, this social dialogue further. In this respect, the interactive, interdisciplinary and innovative approach as proposed by the Taskforce Valuable Agriculture, was directed to a new approach (workshop) to setting up a socially responsible research agenda. In addition, many scientists consider they are not sufficiently equipped for the process aspects. The workshop had positive side-effects on communication between the societal participants. To actively advance the social dialogue, greater effort is needed than was possible in a workshop.

Given the financial and time limitations, the chosen approach led to a reasonably intensive exchange. For the first time, wide structured communication between those involved socially, research representatives, business and government has provided a sound basis for responsible research policy on FMD. All parties contributed basic knowledge and experience, on which the problem analysis and solutions directions were built. The societal participants were in general pleased about the quality of the communication.

Communication in the FMD Workshop

In framework of the FMD Workshop, two workshops were held but stakeholders and scientists did not come into direct contact with one another. This approach was chosen in order to stimulate discussion between the different people involved and to stimulate scientists to work with social priorities and to contribute to the solution pathways as proposed by the stakeholders.

Separate workshops were also chosen because FMD crisis is often an emotionally charged topic for stakeholders, whereas researchers tend to adopt a more rational-analytic approach. Furthermore, there are considerable differences in the expressions and style of communication used by stakeholders and researchers. The social and communicative qualities required of researchers carrying out an integrated FMD research agenda are, therefore, as equally important as their scientific contribution.

Contact was maintained with participants in the stakeholders workshop by inviting several people to attend the scientists workshop and by involving them in preparation of the final report. It is important that stakeholders play an ongoing role in the implementation of the research programme.

It was also decided to restrict the workshops to invited participants and they had the freedom to express their opinions and to reach conclusions. That appears to have been a wise choice but with the disadvantage that communication with WUR colleagues and those further a field occurred later in the form of two press releases.

It also appears that the workshop outcome differs from the outcomes of the separate and standardised interviews conducted by the Agriculture Economics Institute (LEI). This indicates that when different parties were brought together, it is possible to reach a consensus. This can be seen as the added value of the workshop as an approach.

Relationships with the Wageningen UR improved

As a result of the FMD Workshop, contacts have been made between Animal Production Systems, Business Economics Group, Agriculture Economic Institute, ID Lelystad, Alterra, Rural Sociology and Communication & Innovation Studies and between scientists in these groups. This will improve internal and external communications in the future and to greater cooperation between the different groups.

2.7.4 Conclusion about the process

No company representatives at the stakeholders workshop

The list of those invited to attend the stakeholders workshop included representatives of business and industry, such as dairy and feed cooperatives, transport and small and middle-sized companies. Several of these organisations were prepared to have an orientation interview (see Appendix 4) but not to attend the workshop. The workshop participants considered this was a missed opportunity.

No investigation was carried out as to why the representatives of business and industry did not attend. Perhaps, the initial approach of orientation interviews and by letter was not appropriate for this target group. Involvement of this group must be given more attention in a follow-up.

Professional facilitation

It was decided to use external professional facilitation for the workshops because this type of expertise is not available within WUR. Professional facilitation safeguarded the quality and objectiveness of the discussions. In addition, external facilitation gave the organisers more time for reflection during the meeting about the process and the organisation of the workshops.

Financing the FMD Workshop

The budget for the FMD Workshop was € 66,000. For a relatively small amount of money in a very short period with the help of stakeholders and scientists, a great deal of information was obtained about the 2001 FMD crisis. The Animal Production Systems Group provided considerable manpower input. If the time of stakeholders and scientists were included, the workshop cost would have been much higher. A sound financial basis is needed to implement an integrated research programme on FMD.

The workshop methods used are still under development and must be extended in the future. Such Knowledge Workshops are needed to strengthen the role of WUR in society. More financial support for this type of project therefore needs to be made available by the Board of Management and others.

Education

Knowledge gained in the FMD Workshop will be used in the special university course on "Issues and Options in Livestock Production". One of the subjects in the third year degree programme is 'Control of Disease Outbreaks' and MSc students will be able to opt to participate in an integrated FMD research programme.

High demands on leadership of the FMD Workshop

FMD is an emotionally charged subject, with workshop participants often having to defend their positions. This placed heavy demands on personal integrity of the workshop coordinators. The workshop experience indicates that WUR is seen socially as an extension of the Ministry of Agriculture, Nature Conservation and Food Quality, that some scientists from the perspective of their specific discipline think they can solve all FMD related problems, and that communication is governed by emotions and loyalty sensitivities. Good facilitation and acceptance of this emotional aspect of FMD is essential for the success of this type of workshop and for future research.

The organisers experienced the FMD Workshop as a highly successful experiment.

References

Anderson, I., (Chairman), 2002. Foot and Mouth Disease 2001: Lessons to be Learned Inquiry Report, London. www.fmd-lessonslearned.org.uk (15/8/2002)

B&A Groep, 2002. FMD 2001. De Evaluatie van een Crisis. Eindrapport. B&A Groep Beleidsonderzoek & -Advies BV., Den Haag. www.minMinistry of Agriculture, Nature Conservation and Food Quality.nl/infomart/parlemnt/2002/par02080a.pdf (15/8/2002)

Berentsen, P.B.M., A.A. Dijkhuizen and A.J. Oskam, 1990. Foot and Mouth Disease and Export. An economic evaluation of preventive and control strategies for The Netherlands. Wageningen Agricultural University, Wageningen.

Bieleman, 2001. Van Alle Tijden. Trouw zaterdag 7 april 2001.

Boer, J. de, 2002. Het ga je goed, lieve Evelien. FMD-dagboek van een dierenarts. Strengholt. Naarden.

COT, 2002. Besmet gebied. De FMD-crisis getraceerd. Kluwer, Alphen aan den Rijn. COT reeks nr.11.

Curry, D. (Chairman), 2002. Farming & Food. A sustainable future. Report of the policy commission on the future of farming and food. www.cabinet-office.gov.uk/farming

de Bont, C.J.A.M. and J.H. Wisman, 2002. FMD Directe Economische Gevolgen. Studie 2002.Onderdeel van project preventie en bestrijding mond- en klauwzeer, WU-LEI

Denkgroep Wijffels, 2001. Toekomst voor de veehouderij, agenda voor een herontwerp van de sector, denkgroep onder voorzitterschap van de heer H.H.F. Wijffels.

Dierenbescherming, 2002. Evaluatie van het welzijn van dieren tijdens de mond- en klauwzeercrisis in Nederland in 2001(ontwerp). www.dierenbescherming.nl

KNAW, 2002. Bestrijding van mond- en klauwzeer. 'Stamping out', of gebruik maken van wetenschappelijk onderzoek? Koninklijke Nederlandse Akademie van Wetenschappen, Amsterdam.

LTO, 2001. Evaluatie FMD LTO Concern. www.lto.nl/sectoren/rundvee/13december2001/evaluatieFMD.htm (15/08/2002)

The Royal Society, 2002. Infectious Diseases in Livestock. Scientific questions relating to the transmission, prevention and control of epidemic outbreaks of infectious disease in livestock in Great Britain. London, 2002. www.royalsoc.ac.uk/inquiry/index.html (15/08/2002)

Siemes, H., 2001. Ruimen, tien keer erger dan ik dacht. FMD - het boerenverhaal. Elsevier Bedrijfsinformatie, Doetinchem.

Stichting Werkgroep FMD Fryslân, 2001. Tongblier. De Crisis van 2001. Een monument. Stichting Werkgroep FMD Fryslân, Buitenpost.

Taskforce Waardevolle Landbouw, 2001. Naar een WaardeNvolle Landbouw. Wageningen UR, Wageningen.

Van Haaften, E.H. and P.H. Kersten, 2002. Veerkracht. Alterra, Wageningen, Alterra-rapport 539.

Van der Ploeg J.D., A. van Cooten, T. Kierkels and A. Logemann (Ed.), 2002. Kleurrijk Platteland. Zicht op een nieuwe land- en tuinbouw. Koninklijke Van Gorcum bv. Assen.

Websites

www.agriholland.nl/dossiers/FMD/ (15/08/2002)

www.defra.gov.uk/corporate/inquiries/index.asp (15/08/02002)

www.inspection.gc.ca/english/anima/heasan/fad/fmd/sumsome.shtml (15/08/2002)

www.lto.nl/sectoren/rundvee/mondklauw.htm (15/08/2002)

www.minMinistry of Agriculture, Nature Conservation and Food Quality.nl/infomart/dossiers/FMD (15/08/2002)

www.wageningen-ur/FMD (09/09/2002)

www.wageningen-ur.nl/WageningenAteliers/ateliers/atelier_01.html (09/09/2002)

http://www.wau.nl/pers/02/033.html

http://www.wau.nl/pers/02/047.html

www.web.mit.edu/urbanupgrading/upgrading/issues-tools/tools/problem-tree.html (28/08/2002)

Chapter 3. Evaluation of Foot and Mouth Disease policy during the 2001 epidemic in the Netherlands. A project summary

3.1 Introduction

This section reports on a study of policy on Foot and Mouth Disease carried out in 2002 by the Agricultural Economics Research Institute (LEI) and the Business Economics Group (FMA) at Wageningen UR.

The first signs of what was to become an epidemic of Foot and Mouth Disease (FMD) were detected on 19 February 2001 at an abattoir in the United Kingdom. Preventive measures were immediately put in place by both the British government and governments on the European continent to prevent the disease spreading. They involved, for example, a ban on the transport of livestock within the UK, a ban on exports of livestock and meat from the UK to other European countries, and the slaughtering of animals at suspect farms, both in the UK and in countries where British animals had recently been imported. Despite the stringent measures taken, the disease spread to Ireland and France, and from there to the Netherlands. At the start of the epidemic, the control measures involved slaughtering animals at farms where the disease had been detected and at farms that were considered to be at risk, the imposition of protection and monitoring zones, preventive slaughtering of animals on neighbouring farms, and bans on the transport of animals. As the epidemic progressed, additional measures were introduced, including emergency vaccination. Initially, only farms within a radius of 2 km around an infected farm were involved; in early April, however, mass vaccination was carried out of susceptible animals in a triangle between the towns of Apeldoorn, Deventer, and Zwolle.

In order to protect commercial interests, all vaccinated animals and products made from them were later destroyed. This involved killing approximately 270,000 animals, almost 200,000 of them after vaccination. The last animals put down because of the epidemic were slaughtered on 23 May and just over a month later, on 25 June 2001, the Netherlands was officially declared free of Foot and Mouth Disease.

Given the restricted extent and duration of the epidemic, control of the FMD virus was effective from the epidemiological point of view. However, the method of dealing with it raised a large number of issues, with there being a great deal of public unrest in response to the closing off of parts of the country, the killing of large numbers of animals, and the large-scale destruction of carcasses. In addition to the farmers directly involved and the authorities, many other groups of people felt deeply affected by the epidemic. As a result, the outbreak developed into a major political and social crisis, raising fundamental issues regarding policy on controlling FMD and the way that policy had been drawn up.

The Social Studies Knowledge Unit (*Kenniseenheid Maatschappij*) at Wageningen UR also wished to contribute to the discussion and a decision was taken to thoroughly evaluate Dutch policy on FMD. The study covered not only the epidemiological and economic aspects of the FMD crisis but also the psychosocial effects. It dealt with the indicators needed to evaluate various control methods and whether they make possible a balanced decision on the desirable approach to control so as to minimise any crisis resulting from future outbreaks. These were the central issues dealt with in the study.

3.2 Aim of the study

The aim of the study was to make a contribution to future policy on tackling Foot and Mouth Disease in the Netherlands and in Europe by carrying out an integrated socio-economic analysis of the problems posed by this disease. The concrete objectives of the project were:
- to determine the qualitative (social psychology) impact of the Foot and Mouth Disease epidemic that occurred in the Netherlands in 2001;
- to determine the quantitative (economic) impact of the epidemic;
- to carry out an epidemiological and economic foresight study and evaluation of the policy options for combating any future outbreak of FMD;
- to produce an integrated evaluation framework for assessing policy options.

3.3 Structure of the research project

Parallel with the four objectives, the research project consisted of four subprojects. The relationship between them is shown in Figure 1. The qualitative and quantitative evaluation of the 2001 epidemic (subprojects 1 and 2, respectively) provided information on the impact of the epidemic on various groups. After the crisis, the authorities, politicians, and industry realised that in future these groups needed to be given a greater say in determining policy. Within the context of this project, that intention was given shape in subproject 4. The quantitative evaluation also produced information on the economic losses; this was used in developing and applying economic model calculations (subproject 3). Together with the psychosocial effects, the epidemiological and economic consequences of new policy that were calculated (subproject 3) provided input for the sociological evaluation (subproject 4).

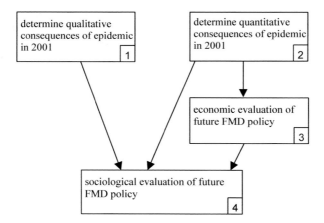

Figure 1. Relationship between subprojects.

As already mentioned, this document deals specifically with the qualitative impact of the FMD crisis and the way in which future policy on the disease could be made far more interactive (governance), with account being taken of the interests of all the stakeholders involved (the authorities, industry, and various other organisations).

3.4 Qualitative impact of the 2001 outbreak of FMD

3.4.1 Introduction and approach

This subproject focused on qualitative analysis of the FMD crisis, considering the subjective perception and experience of various different groups. It applied a stakeholder analysis, evaluation of the communication regarding FMD, and a perception/experience study of the population. The stakeholder analysis was intended to clarify the interests, points of view, and motives regarding FMD among the various different stakeholders in the Netherlands. Evaluation of the communication looked at stakeholders' communication strategy, also investigating their risk management and risk assessment procedures. The perception/experience study was intended to determine how the public experienced the FMD crisis. This subproject focused specifically on exploring the standards and values that were involved.

3.4.2 Stakeholder analysis

A stakeholder analysis was carried out as part of the qualitative analysis of the impact of the FMD crisis. Its objective was to clarify the interests, points of view, and motives underlying attitudes to the crisis and its effects among the various different stakeholders in the Netherlands. The analysis was based mainly on in-depth interviews with a large number of those concerned.

The qualitative effects of the FMD crisis are discussed here with respect to three different periods: (1) the period prior to the epidemic, running up to the official announcement on 21 March 2001; (2) the period of the actual epidemic (from 21 March 2001 on); and (3) the period after final screening op 25 June 2001 (the post-crisis period).

This approach is based on the causality of the crisis. The underlying thinking is that if neither the authorities, the agricultural sector, nor the various other relevant organisations are properly prepared for an epidemic of animal disease, there is a major risk of such an epidemic leading to a large-scale crisis.

The main conclusions regarding the period prior to the FMD crisis were:
- During this period, it was primarily the short-term outlook of a large number of stakeholders that played a role and their actions were primarily determined by operation-specific economic factors.
- There was little enthusiasm among the stakeholders for a jointly determined strategy.
- Coordination by the Ministry of Agriculture, Nature Management and Fisheries did not inspire commitment.
- A number of stakeholders had little confidence in the Ministry's way of tackling FMD.
- The Ministry was seen as insufficiently open to effectuate a joint approach.
- There was little or no collaboration between different sectors in dealing with a potential FMD crisis. BSE was viewed as a more significant danger and was tackled within individual sectors and organisations. FMD did not really come into the picture.

The main conclusions regarding the period of the actual epidemic were:
- A large number of organisations found it difficult to set up a crisis team during the first few hours after the epidemic was announced.

- There was little or no direct collaboration at national level between the various stakeholders during the crisis.
- There was in fact effective collaboration in dealing with the problems at regional government level, particularly the municipalities in the triangle formed by the towns of Apeldoorn, Zwolle, and Deventer.
- The Ministry's scenario failed to stress the value of a joint strategy. As a result, stakeholders relied solely on themselves and their own contacts and in many cases operated independently, taking decisions on an ad hoc basis.
- In tackling the crisis, the government took very little account of the emotional aspects involved when families have kept animals for generations or with feelings in the hotel and restaurant trade. This led to a great deal of opposition.

The main conclusions regarding the post-crisis period were:
- The way the crisis was handled from the financial and organisational point of view led to a great deal of dissatisfaction in a number of sectors. Businesses were left in a state of uncertainty regarding compensation for far too long.
- Those in the transport sector considered that they had been wrongly accused of being partly responsible for the epidemic.
- The epidemic and the ensuing crisis led to a shift in thinking regarding the way FMD should be tackled. Various stakeholders (particularly relevant associations and interest groups) opposed the killing of large numbers of healthy animals.
- The Ministry of Agriculture, Nature Management and Fisheries seemed to only gradually realise that preventing and combating FMD is an issue affecting the overall public interest and that an interactive approach that involves stakeholders is essential in creating a basis of support for the action taken and in preventing public disquiet.

3.4.3 Evaluation of communication regarding FMD

The stakeholder analysis also showed that a large number of communication problems arose on the side of the stakeholders. This involved communication between the various stakeholders on the one hand and between the stakeholders and those they represent on the other. With that in mind, it was important to understand how the parties communicated with one another about the FMD crisis. The analysis was primarily based on in-depth interviews with a number of the stakeholders concerned. One can draw the following conclusions regarding communication before, during, and after the FMD crisis:
- When the epidemic broke out, most organisations were unprepared for dealing with it and had no formal communication policy in this area. Even so, many organisations quickly set up a crisis team which also became involved in communication.
- Stakeholders with similar interests sought out one another depending on their needs, the actual matter concerned, and the associated interests. Although this did not produce any cohesive consultation structure between them, there were very few communication problems between the various stakeholders.
- Most of the communication problems that stakeholders had were with the Ministry of Agriculture, Nature Management and Fisheries, particularly regarding the speed of communication on the part of the Ministry and the limited degree of empathy shown by the Ministry with respect to the practicalities faced by the other stakeholders. In a number of cases, this had a negative effect on relations between the Ministry and other stakeholders. A number of stakeholders have indicated that the strictly legalistic approach to

communication adopted by the Ministry means they will be unlikely to collaborate with the Ministry in dealing with any future crisis, except on certain conditions.

3.4.4 Perception/experience study of population and consumers

The 2001 FMD crisis brought about a fundamental reconsideration of methods of preventing and combating the disease. It also led to a great deal of public disquiet and criticism of the policy pursued by the authorities. In reviewing the crisis, one therefore needs to consider the grounds for this criticism and how people now look back on what happened.

The study involved analysing the way farmers and the general public perceived the crisis. The analysis itself involved a quantitative study of a large number of people. A questionnaire was drawn up and presented to the Internet panel used by the firm of Motivaction. This consists of more than 17,000 people representing the 15 to 55 age group of the Dutch population. The questionnaire comprised approximately 100 statements regarding standards and values and a large number of questions about attitudes and behaviour in specific areas.

This subproject allows one to draw the following conclusions regarding the way farmers and the general public perceived and experienced the FMD crisis:
- The FMD crisis made a great impression on people, both rationally (the approach adopted) and emotionally (greater concern for animals than is sometimes thought). A large number of those involved are still dealing with the aftermath of the crisis.
- A significant proportion of people realise that animal diseases are a natural occurrence but they also think that action should be taken when an outbreak is serious.
- In general, people believe that human beings are part of nature and are not superior to it.
- People have their doubts about the safety of products produced from vaccinated animals.
- People support a policy of vaccination; farmers are also concerned about what this could mean for exports.
- Farmers are divided as to whether animals or the economy should be the main factor. This also affects the way farmers as a whole view FMD control measures.

3.5. Towards a framework for sociological evaluation

3.5.1 Introduction

In this section, we define an integrated framework for evaluation that will allow us to compare all the relevant aspects of various different strategies for dealing with FMD, including their public impact. This framework specifically allows for the fact that not all the relevant factors can be expressed in monetary terms. This applies particularly to a number of psychosocial and ethical factors that are of great importance, as shown by the qualitative analysis of the effects of the 2001 FMD epidemic (see above).

3.5.2 Multi-Criteria Analysis

To develop the evaluation framework, the research project described here made use of Multi-Criteria Analysis (MCA). MCA is a general method of approaching problems that involve choosing between a number of alternatives. It makes it possible to select the "best" alternatives or to rank alternatives in order. MCA starts by determining what alternatives are available and formulating the criteria for assessing them. The alternatives are then scored according to those criteria.

The next step is to determine the relative importance of the criteria, i.e. the "weighting" of each of them. The weightings assigned may be quantitative but in many cases it is more realistic to assign weightings on a qualitative basis.

The scores for the various criteria are then linked to the weightings. This study makes use of the "weighted summation method", in which the standardised criteria scores are multiplied by the corresponding weightings and then added up for each alternative. This produces a total score for each alternative, allowing them to be ranked in order.

The criteria, indicators, and weightings were selected by a steering committee consisting of representatives of the authorities, industry, and various other organisations. A session was held – in a "group decision room" (GDR) – at which the steering committee set the basic criteria, indicators, and weightings for an integrated evaluation framework.

Performing an MCA is basically subjective because the analysis is based on the preferences and perceptions of those concerned and these are built into the model. The result of the MCA therefore depends on those involved in constructing the model. If this method is to indicate the most appropriate method of dealing with FMD – the method enjoying the greatest degree of public support – it is therefore necessary to repeat the MCA periodically with representatives of the main stakeholders. The authorities should then base their policy on the results of this procedure with a view to practising the most acceptable form of governance. Here, however, we will confine ourselves to the results of the analysis by experts.

3.6 Results of the Multi-Criteria Analysis

The experts selected nine criteria on which to base the integrated evaluation of FMD policy. They are summarised in Table 1 in order of the average weighting (maximum = 100%) assigned to them by participants in the GDR session. In the MCA, each of the criteria is further subdivided into subcriteria so as to be able to accurately determine the scores for each criterion and for each policy alternative (i.e. method of combating FMD).

As an illustration, this section examines the evaluation in which the integration framework was applied so as to produce a "weighted ranking" of the various different strategies for dealing with FMD epidemics in a given region. Evaluation was based on the following options (EU = EU basic package, 72NSS = 72-hour national standstill, Slaughter = ring slaughtering, Vac = ring vaccination, slaughter/preserve = slaughter vaccinated animals or preserve them):
- Slaughter1: EU + 72NSS + Slaughter_1km
- Slaughter2: EU + 72NSS + Slaughter_2km
- Vac1:   EU + 72NSS + Vac_1km (slaughter/preserve)
- Vac2:   EU + 72NSS + Vac_2km (slaughter/preserve)
- Vac4:   EU + 72NSS + Vac_4km (slaughter/preserve)
- Slaughter1+Vac2: EU + 72NSS + Slaughter_1km + Vac_2km (slaughter/preserve)

Table 1. Selected criteria for integrated evaluation of an outbreak of FMD and the weightings assigned to each criterion by nine experts.

Criterion no.	Specification	Weighting (%)
1	International feasibility: the likelihood that the strategy will be supported in the international contexts within which the Netherlands has obligations	20.0
2	Epidemiology: the extent to which a strategy can contribute to eradicating an infectious disease of animals	19.4
3	Macro-economics: the effect on national income and the balance of payments	17.8
4	Ethics: animal welfare	12.8
5	Micro-economics: the distribution of the effects across various different sectors and regions	11.7
6	Psychology: traumas affecting those involved	7.2
7	Economics (international): consequences of action taken by the Netherlands for other countries	5.5
8	Ecology: effects on the environment, nature, and biodiversity	2.8
9	Inconvenience: disruption of normal activities caused by restrictions on movement (this is separate from any loss of income resulting from this inconvenience)	2.8

Converted into scores for each alternative, the results of the analysis by experts are as follows:

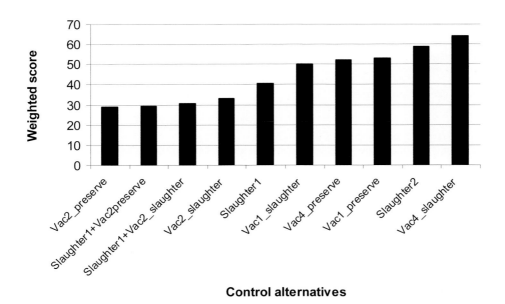

Figure 2. Total scores for each alternative.

Table 2 shows the results of the weighted score for each criterion and control alternative. *The higher the score, the more negative the judgment.* The "Slaughter_2km" alternative, for example, does not get a high score according to the criteria Epidemiology (score 18.8) and Ethics (score 13.15). The maximum possible total score for each option is 100. Based on this analysis, the total score for the various different strategies ranges from 29.18 to 64.10. The best strategies are Vac_2km and Slaughter_1km + Vac_2km, both when vaccinated animals are slaughtered and when they are preserved. Vac_4km_slaughter and Slaughter_2km are the worst options. For a graph of the various different rankings, see Figure 2.

Table 2. Weighted score for each criterion and control option (preserved = vaccinated animals are preserved, slaughter = vaccinated animals are slaughtered).

	1[a]	2	3	4	5	6	7	8	9	10
Epidemiology	9.69	18.80	14.12	3.82	11.85	2.12	15.02	6.35	16.26	3.68
Macro-economics	7.15	6.53	15.83	12.24	15.87	10.15	9.05	6.24	11.29	3.88
Ethics	8.39	13.15	5.97	1.36	4.15	4.65	8.57	6.00	10.33	8.28
Micro-economics	4.03	5.71	5.10	4.31	5.94	3.98	6.50	6.77	11.37	6.50
Emotions	5.53	7.12	2.91	1.46	4.43	2.67	3.20	2.04	5.72	3.07
Int. econ. consequences	1.86	2.14	4.97	4.42	4.84	4.18	3.83	3.28	3.71	3.04
Ecology	1.31	2.27	2.20	1.05	1.82	0.98	2.33	1.57	2.55	1.27
Nuisance	2.33	2.89	1.55	0.53	2.88	0.67	1.55	0.53	2.88	0.67
Total score	**40.29**	**58.61**	**52.64**	**29.18**	**51.78**	**29.39**	**50.05**	**32.79**	**64.10**	**30.38**

[a] 1 = Slaughter1 2 = Slaughter2 3 = Vac1_ preserved
4 = Vac2_ preserved 5 = Vac4_ preserved 6 = Slaughter1 +Vac2 preserved
7 = Vac1_ slaughter 8 = Vac2_ slaughter 9 = Vac4_ slaughter
10 = Slaughter1 +Vac2_ Slaughter

Summarising, the conclusions of the MCA are:
- The MCA method used makes it possible to develop a tool for determining, in a structured manner, the conflicting interests of the various stakeholders and assigning a weighting to each control option. A tool of this kind can therefore facilitate consultation and discussion regarding such complex, integrated decisions as the choice of a strategy to combat FMD.
- The relevant public attitudes towards methods of combating FMD vary over time. Creating an evaluation framework is therefore a continuous process. The methodology used is an extremely flexible one, making it possible to update indicators or weighting factors without too much difficulty.

3.7 Overall conclusions

Our main conclusions are as follows:
- Although the extent of the 2001 outbreak of FMD was restricted from a veterinary point of view, it had a far-reaching social, psychological, and ethical impact, both within the agricultural sector and elsewhere. This was partly due to the communication policy pursued by the authorities and industry, which was considered to be faulty.
- It is important for the authorities to determine strategies for dealing with any future FMD epidemic in consultation and collaboration with the stakeholders involved. Govern<u>ance</u> (interactive policy-making) is a better strategy in this respect than govern<u>ment</u>.
- Viewed in the context of a wide-ranging evaluation framework (one that includes psychosocial factors), strategies involving regional emergency vaccination (with relatively small vaccination areas) score highest. The larger the number of animals that need to be slaughtered (as part of a policy of preventive slaughter or vaccination), the lower the score.

Appendix 1. Causes and effects of the FMD outbreak and crisis

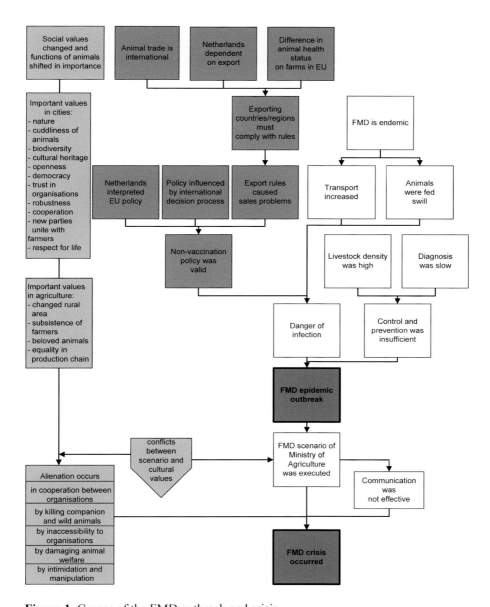

Figure 1. Causes of the FMD outbreak and crisis.

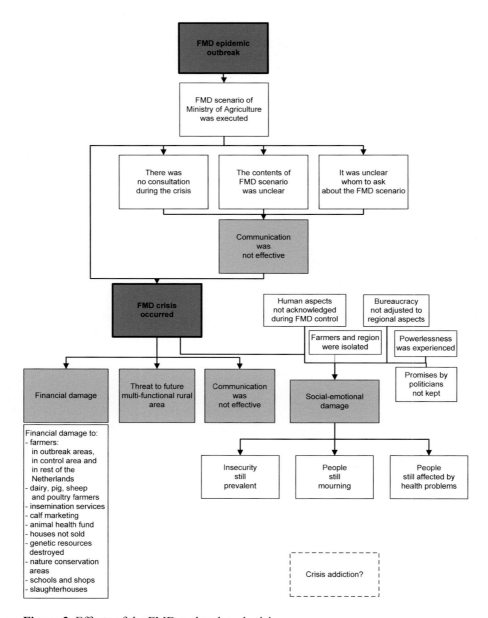

Figure 2. Effects of the FMD outbreak and crisis.

Appendix 2. Report of workshop 1: Workshop with stakeholders held 18 April 2002

1. Aim of the workshop

The aim of this workshop was to arrive at a joint analysis of the FMD policy through discussion of the various aspects of the problem and exchange of experiences. Thus far the problem had only been discussed separately among the groups and not en bloc. This analysis would then form the keystone for the search for ways and means to prevent new outbreaks and a new alternative approach should a new outbreak occur. Such an approach would involve a survey of the techniques and financing but also of the underlying values of humans and animals. Discussion of the criteria for assessing solution pathways would follow and serve to determine the priorities of the solution pathways.

2. Method

2.1 Target group:

Of the 57 people invited to participate in this workshop 25 attended. These were people who were involved in the FMD outbreak in 2001 spread across the private sector, the primary livestock husbandry sector, government and social organisations. All of these parties were represented at the workshop.

2.2 Set up of the workshop

The day began with a presentation of a problem tree derived from the interviews with the parties involved, together with an explanation of the principle of a problem tree. Then the main problem was tackled: in this case the FMD epidemic. Causes and effects of the problem were listed and ranked, after which problem groups (with causes and effects) were bundled together to form the branches.

Following the presentation of the problem tree the participants were divided into 4 groups to further discuss one branch of the problem tree each. The other groups that were formed were: communication, socio-emotional damage, multifunctional rural areas and social values, international policy and financial consequences.

Plenary presentations of the results of the working groups followed in relation to the problem tree as a whole.

The second part of the workshop consisted of a brainstorm session on possible solutions for the branches. The effects of these solutions were examined in the light of the problem tree as a whole.

The day was closed with a discussion on the criteria that the solution pathways would have to satisfy. The priorities of the suggested solution pathways were geared to the criteria.

3. Results

In the four subgroups that discussed the different branches of the problem tree the following came up:

3.1 Communication

Three groups played a role in the communication surrounding the scenario, namely, the Ministry of Agriculture, Nature Management and Fisheries (LNV), farmers and the other parties involved – those involved with animals in zoos and hobby farms and those without animals such as mid-sized and smallholdings and catering establishments. The influence of the media had a major role to play in communications surrounding the FMD crisis.

Communication on the part of the Ministry of LNV during the FMD crisis was mainly informative in nature. There was room for discussion in the basic consultation prior to the crisis. The content of the scenario and the reasons underlying the rules set down in it were unknown to the producers and involved parties, which made the communication process unclear. The performance of the Minister of LNV (Brinkhorst) led to the hostile attitude that many parties adopted towards the LNV.

Furthermore, there were problems about the manner in which farmers were notified about an infection on their farm: when a veterinarian suspected the presence of an infectious animal disease on a farm, samples were sent to the laboratory. If there was no infection the farmer was told this by his own veterinarian. If there was indeed an infection on the farm the veterinarian was not told and the farmer was notified of it by the RVV (Meat and Livestock Inspectorate). This last-mentioned route was experienced as very standoffish.

The most important cause-effect relations about communication are presented in Figure 1.

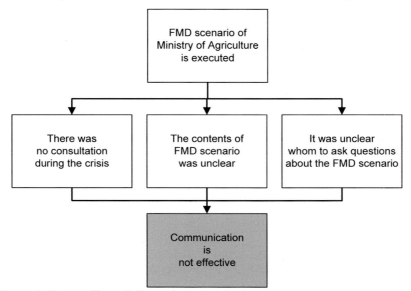

Figure 1. Cause-effect relations of communication.

3.2 Socio-emotional damage

The key question that this group tackled was: why did the crisis hit so hard? The reasons mentioned were (Figure 2):
- Inexplicability of the death of huge numbers of healthy animals.
- The way in which people were treated was disrespectful at times.
- The FMD outbreak was seen as a veterinary problem. In the solution to the problem the human factor was not taken into account.
- The farmers felt that the Minister of LNV's remarks to them were negative in character implying that they were fussing about nothing and that they should stop complaining. This incited much incomprehension.
- Isolation of the farmers during the crisis deeply affected them and their well-being. They felt as though they too were treated as infectious.
- In addition, many farmers were forced to undergo developments and were powerless. They were not allowed to participate in the control of the crisis and weren't even in charge on their own property.
- The control of the crisis was organised far too centrally. Implementation problems could have been avoided by calling on local expertise.
- Bureaucracy functions well in peacetime, but falls short in wartime.

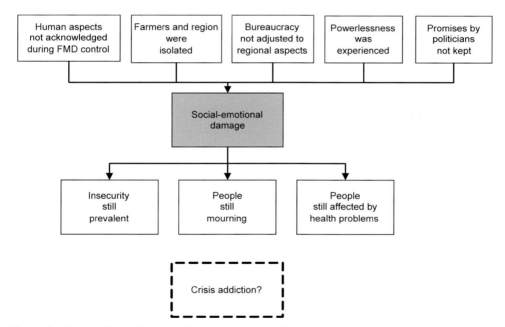

Figure 2. Cause-effect relations of social-emotional damage.

3.3 Multifunctional rural areas and shifts in social values

Social values have changed since the scenario was drawn up in 1990. In addition, an apparent difference has occurred between the values in the city pertaining to the citizens and the values of agriculture pertaining to farmers, causing a chasm between the city and the rural area. It has led to the recognition of the fact that the values in the entire problem tree with all the causes and effects are intertwined and that the societal values must therefore come high up in the problem tree (Figure 3).

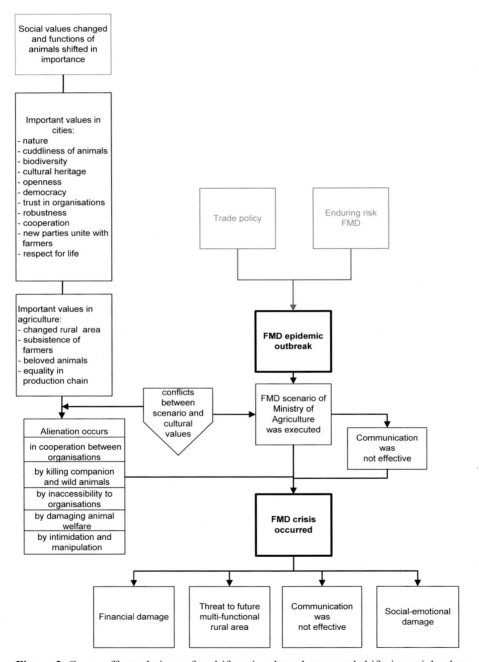

Figure 3. Cause-effect relations of multifunctional rural areas and shifts in social values.

3.4 International policy and financial consequences

This group has been added on to this branch of the problem tree as Figure 4 shows.

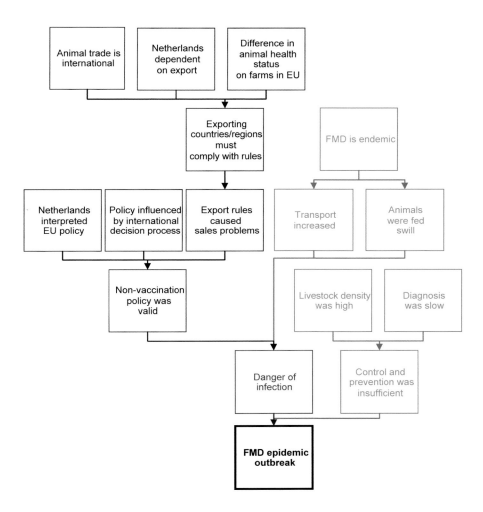

Figure 4. International policy component of problem tree.

The financial consequences that affected this group have been summed up and divided into direct and indirect costs. The direct costs were:

In the vaccination area:
- depletion of farmers' incomes (differed per sector/farmer/region)
- no income for culling and starting up problems with new livestock

In the control area:
- decrease in farmhouse sales
- decline in second income
- loss of breeding lines
- financial losses in many sectors

Across the Netherlands:
- costs for site managers to close terrains

The following groups suffered indirect costs:
- slaughterhouses
- (animal) transport
- recreation
- children's farms
- hobby farmers
- shops
- international market position of animal husbandry in the Netherlands weakened

The indirect costs for humans:
- people ill as a result of stress
- children forced to repeat a year of school
- less spending power

4 Results of the afternoon discussion

In the afternoon discussion the participants had a brainstorm session on solution pathways covering the same topics as the morning discussion.

4.1 Communication

- A 'complete' scenario should be drawn up starting from the present culling arrangements, which should also include the evaluation of this crisis as well as the communication process.
- The scenario must be better communicated in 'peacetime' to those involved as well as the producers. Information to these groups must be proactive. And during a crisis these groups should be able to approach local authorities with their problems who in turn should be able to convert them into local circumstances.
- During a crisis information from a region must be fed back to central government.
- International communication needs to be improved especially for knowledge enhancement. Policy change must be coordinated by a neutral organisation.

4.2 Socio-emotional damage

- The crisis organisation must be local (bottom-up), whereby there is coordination between departments and LNV and agreement between LNV and the local environment via the regional direction of LNV. Local practical expertise is required to be made use of including veterinarians, police, clergy and local authorities. A pilot study of this local scenario needs to be done and evaluated. It is important that the local experience is respected and that competencies are clear.
- The human factor needs to be made obvious, by appointing a local informant such as the veterinarian, for instance. Apart from this the government should devote more attention to the human factor through recognition, compassion and visiting the stricken regions. Members of the government should adopt a respectful attitude when approaching the affected regions via the media.
- The feeling of powerlessness must be prevented. Clear information and an accessible government can help to achieve this.

4.3 Multifunctional rural area

- An independent institute (within WUR) needs to be installed where the continuity of the FMD workshop can be safeguarded.
- A new interactive, region-oriented scenario needs to be drawn up. And we need to start with a clean slate.

4.4 Trade politics

- Formulation of the scenario.
- The Netherlands and the European Union should have the selfsame policy.
- There should be stricter border control and if suspicions are aroused borders must be closed promptly.
- The meat and dairy processing chains must be held responsible for the processing of products from vaccinated animals.

There is overlap in the solution pathways such as suggested by the different discussion groups. In the plenary discussion on this issue, a new solution pathway was added:
Replacing the reduction of compensation with fines.

4.5 Criteria

The solution pathways must be tested according to several criteria to determine which of the solution pathways should be given priority. The criteria formulated by the parties involved are:
- Support of solutions in society
- Feasibility of the solution: from a technical, process and legal point of view.
- Costs of efficiency of solution (financial)
- Results of the solution for the different sectors such as nature, recreation, middle and small holdings

- Sustainability of the solution
- Ethical soundness of the solution
- Immaterial damage caused by the solution

Based on the above criteria the participants of the workshop voted for two of the solution pathways that they gave priority to from where the following solution pathways were derived:
- There has to be a completely new scenario (16 votes)
- The EU framework must be discussed (12 votes)
- A pilot scenario should be carried out in one region (7 votes)
- The food chain for the processing of vaccinated animals must be addressed (7 votes)
- Converting the discount system to a system of fines (4 votes)
- Stricter border control and earlier closing of the borders if FMD is suspected (0 votes)

5. General conclusions

This extended problem tree and the four most important solution pathways that emerged from this workshop will form the launching pad for the second workshop, which will be held with researchers who are involved with different aspects of FMD.

Appendix 3.   Report of workshop 2: Workshop with scientists held 5 June 2002

1. Aim of the workshop

The aim of this workshop was to bring together researchers working in the field of FMD in order to make an inventory of current research, identify topics that are not being investigated and give shape and form to the solution pathways put forward by those who took part in workshop 1.

2. Method

2.1 Target group

Among the scientists invited to this workshop were economists, epidemiologists, veterinarians and sociologists. Also invited to the second workshop were several representatives from the the first workshop. There were 25 participants at the second workshop.

2.2 Programme

The programme commenced with an outline of the day's objectives and an explanation of the principle of the problem tree. The complete FMD problem tree as established during the first workshop was then presented and participants of the second workshop were asked to place his or her research in this problem tree. This created an overview of the ongoing research on FMD. The reseachers were given a maximum of three cards with which they could indicate the fields of their current research, thereby clearly showing up the research fields that were not or barely covered (white space/gaps).

After lunch the group was divided into four smaller working groups. Participants in the working groups discussed the three solution pathways that emerged from the first workshop and the white space that was visible in the morning session of this second workshop. The aim was to give more shape and form to these proposed solution pathways by means of concrete research questions and research fields and to pinpoint the expertise necessary to carry this out. Trajectories and conditions could also be mentioned if needed.

The solution pathways that emerged from the first workshop were:
1. Compiling of a totally new scenario (starting with a clean slate), with a local pilot study to facilitate the process, if necessary.
2. Addressing the food chain (meat, dairy products) on the processing of products from vaccinated animals.
3. Opening up a debate within the EU framework on legislation governing control of FMD.

The day was brought to a close with a brief first reaction from Mr A. Dijkhuizen, chairman of the Executive Board of Wageningen UR, to the results of the discussions that were presented at the end of the day.

3. Results from the morning session

After the presentation of the FMD problem tree there was a discussion about the appropriateness of the problem tree. Among the suggestions was the idea that the entire FMD problem ought to examined more fundamentally. Moreover, it was suggested that the problem tree has been drawn up from the perspective of the affected regions and did not elucidate the perspective of the Netherlands or Europe. The reactive character of the scenario was put forward as a likely extra element of the problem tree and that transactions should be proactive before an outbreak occurs.

Then the research of the different participants was placed in the appropriate spot in the FMD problem tree. One by one the participants introduced themselves to the group, placed their research in the tree (Figure 1) and gave a brief explanation of the research they had done.

Five different colours helped distinguish the researchers, the colour of the card indicating the discipline of each of the participants as follows: Economics (purple), Veterinary Medicine and Epidemiology (green), Sociology (pale green) and the remainder (blue) among whom were tutors and researchers from the RIVM. Yellow cards set apart the representatives from the first workshop.

It was apparent that there was a concentration of economists and veterinarians/ epidemiologists in the scenario, whereas sociologists were involved in research spread across the problem tree. Cautious conclusions were drawn from the placing and spread of the cards and the diverse colours, because not all groups were represented and some groups were bigger than others. The aim was to identify the white space (gaps) in the research and this was achieved. According to the participants more sociologists in the group would not have produced any cards in new places. The participants concluded that the overview of the research and the white space that became apparent as a result corresponded with their expertise in their field of research.

Another conclusion was that the research mainly focuses on symptoms and no attention is paid to fundamental research. It was said that there are no funds for fundamental research and that money for research would only be made available if and when a problem arises.

3.1 Analysis of white space (gaps)

Next, the fields in which little or no research is carried out (white space or gaps) in the FMD problem tree were named. A large number of possible white space was named and summarised in seven white spaces as follows:
1. International framework with questions such as: How autonomous is the Netherlands within Europe? To what extent can the Netherlands interpret EU legislation.
2. Role of animal husbandry in the Netherlands. A multifunctional rural area belongs here.
3. Attention to food chain and consumers and the impact that this chain and consumers have.
4. Social/economic cost-benefit analysis.
5. Interaction between the different scientific disciplines, especially communication.
6. Process of preventive (vaccination) instead of curative control of an epidemic. Components are: control of an outbreak, use of guidelines instead of inflexible scenarios and translation of policy into practical processes.
7. Abstract nature of the process of governing and management decision models, trading methods and style of a crisis.

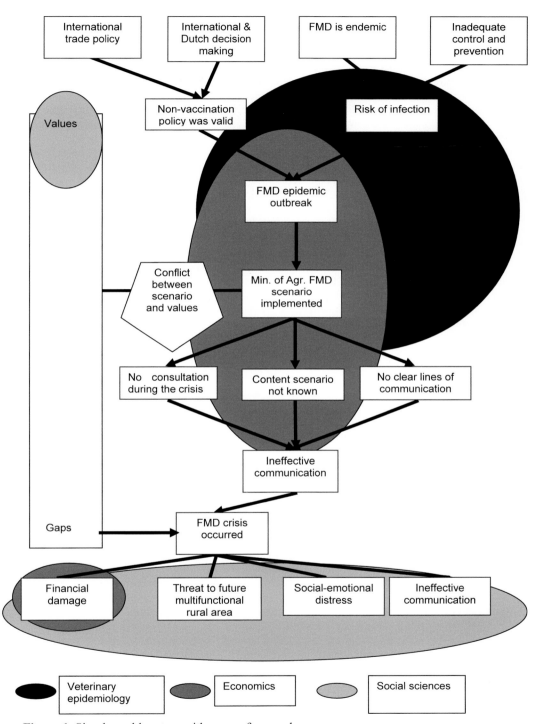

Figure 1. Simple problem tree with areas of research.

ad 1. The international framework was dealt with in a solution pathway formulated by the first workshop to bring up the legislation governing control of FMD for discussion within the EU framework.

ad 2. The role of animal husbandry is a wider discussion that is currently taking place in the Netherlands, broader than what was included in the FMD workshop.

ad 3. Attention to the food chain and the consumer and the impact of chain and consumer was tackled in a solution pathway that came out of the first workshop. This solution pathway dealt with addressing of the food chain (meat, dairy produce) on the processing of products form vaccinated animals.

ad 4. During the discussion on white space (gaps) one participant said that there already was ongoing research on this subject. It implied that there was no white space any longer and therefore the subject was ignored.

ad 5. Interaction between the different scientific disciplines will be a component of the follow-up trajectory of this workshop on FMD expertise. As the Executive Board will be dealing with this phase it wasn't dealt with here.

ad 6. Preventive instead of curative control of a crisis was included in a solution pathway listed in the first workshop. This solution pathway has to do with a totally new scenario (starting afresh) with a local pilot study if deemed necessary for the purpose.

ad 7. This white space (gap) emphasised three dimensions in processes: a content dimension, an organisational dimension and a subjective emotional dimension. The topic was discussed in the afternoon among a group from which it appeared that a scientific approach was necessary to link up these three dimensions.

4. Results of the afternoon discussion

4.1 Group 1

Compiling a totally new scenario (starting with a clean slate), with a local pilot study to facilitate the process, if necessary (Figure 2).

Objective: The involvement of the different stakeholders must be brought to the fore. Support must be created. The stakeholders need to know what to expect at a next crisis and when the operations will be set in motion. At any rate they must have an idea of the content of the scenario. A trial run of the scenario should take place in a certain region.

The approach in the new scenario should be bottom-up taking others into account. There is a need for different scenarios for different types of farms such as hobby farms, experimental farms, dairy farms and so forth. The different groups of stakeholders should be determined via an environmental analysis together with interviews with stakeholders in different ways: via a social debate (for example, biotechnology), via a workshop, the Internet or a public hearing (such as in the US). An inventory must be made of the different methods from which a good one can be selected. Questions consequently asked are: How do you determine who a stakeholder is, how do you prepare stakeholders and how do you achieve a consensus? Participative research as often done in development sciences and communication sciences could be used to arouse the interest of the different. There would still be a difference between farmers and the national consensus but no attempt should be made to put them together. The stakeholders would need to be prepared for the consequences of the different options (vaccination, nonvaccination) for the region, the Netherlands and the EU.

The current emergency vaccination plan, in compliance with EU legislation, has being taken as starting point for drawing up the scenario. Next the different solution pathways will need to be assessed taking into account emergency vaccination and loss of FMD-free status, export and export bans and the possibility of processing and sales inside the Netherlands. The stakeholders must make a choice, bottom-up, from these different solution pathways and the scenario made known to all the different stakeholders.

Communication of the scenarios is important, not only the moment a crisis occurs but also beforehand, for example, through brochures containing information about the scenario made for the different target groups (farmers, scientists and so forth). The media has had much influence on communication during the crisis. The media should be encouraged to also channel their efforts into disseminating information before the crisis arises.

Research questions	Expertise	Trajectory/Conditions
1. Environmental analysis * Who are the stakeholders? * How to get them involved? * Preparation? * Achieve consensus among stakeholders?	Business administration Communication sciences Sociology	Pilot Independent process leader

2. Assuming emergency vaccination:
* Which scenarios are feasible?
* Evaluation among stakeholders
* Economic, social-
  public
  cost-benefit analysis
	FMA, LEI, APS + lots more	Institutional, no squabbling

3. Improve communication * rolls, actors? Aim of scenario? (For EU/ Regions)	Applied research, media- influence Public organisations Design methods	Democratic decision

Figure 2. Matrix group 1.

Comments on the presentation

Involvement of the different stakeholders can take place in various ways as already mentioned in the group discussions. In the Netherlands there are always a lot of stakeholders, which is why special methods need to be developed for the Netherlands because it is difficult to work with such large numbers.

It was also said that stakeholders are not always willing to participate.

Use needs to be made of a different method of decisionmaking: the multi-stakeholder reflex. Although this does not come up with answers, it does come up with designs to reach solutions.

The citizen turns away from the government, whereas previously he was pinned to them. The workshop contributes to the creation of societal support, which in turn contributes towards political decisionmaking. The government is responsible for the policy.

It emerged from workshop 1 that the regional aspect is important within a national framework. For this purpose an independent process leader will be needed plus the accompanying preconditions. The scenario should now be submitted to Brussels, but who should do this? There should be a social scenario that also takes the public at large into account.

4.2 Group 2

Addressing the food chain (meat, dairy products) on the processing of products from vaccinated animals (Figure 3).

Research question	Expertise	Trajectory/Conditions
1. Commercial viability - NL/EU legislation? - Consumer acceptance? - Information	LNV/Industry/retail WUR, Consumer Association	SCALE/TIME? QUESTIONS + INTERNATIONAL
2. Food chain guarantee systems - Tracking en tracing (I&R,…) - Vaccination possibilities diagnosis - Risk assessment (carriers) - Control	LNV+GD+Industry WUR-ID/Animal Expertise Group FMA+QVE+FD(UU), Alterra LNV+WUR+Industry	
Domestic market production Form new public organisation - social undertaking - globalisation - international competition Science also listening to farmers' ideas: top and field => representation, support, LTO,….	LNV, LEI, Consumer Association, LTO, other farmers government, consumer	

Figure 3. Matrix group 2.

In the group it was decided that this concerns the processing of products of vaccinated animals inside the Netherlands. Would the acceptance of vaccinated meat technically, socially and commercially be attainable? Two differentiable systems can be identified here. Following vaccination the animals are slaughted and the meat is sold or the cattle are not slaughtered and trading takes place later. Assuming the latter possibility, a food chain guarantee system would be necessary to indicate whether or not the meat comes from the vaccinated region. To be able to do this an indicator system will be necessary like they use in forestry, for instance. This can serve as a model for a food chain guarantee system for vaccinated meat. Another possibility is a tracking and tracing system based on UBN numbers. The system based on I&R is not conclusive. These systems fall under the Ministry of LNV's responsibilities, whereby European legislation and the animal disease status of a country need to be taken into account. The marker vaccine is necessary to differentiate between vaccinated and infected animals.

It is important to determine how consumers cope with vaccinated meat. Eating vaccinated meat is not a problem, but the existence of trade restrictions does little to boost consumer confidance. It is therefore important to provide the consumer with the proper information. There should be a policy to enforce the consumption of vaccinated meat or the consumer should demand it.

A totally different approach of animal husbandry is a possibility that was proposed, a kind of multifunctional agriculture, in which livestock husbandry moves towards a situation of socially acceptable agriculture that only produces for the domestic market. This would then result in about half of the present agriculture having to disappear and was therefore not seen as a viable alternative by the group.

The LTO (farmers organisation) was not perceived as a representative for the entire agricultural sector, but only representing some of the stronger sectors. LTO cannot be seen as the official voice for the farmers.

It was noted that the public sector was absent at this workshop.

Comments on the presentation

There is a traditional cheese making business in the north of the Netherlands that does not work through the existing cooperatives. Now that this group is growing they are beginning to organise themselves and new cooperatives are starting.

To determine the type of research that they wish to do researchers are forced to listen to financiers.

LNV also envisages a role for the food chain guarantee system.

Can we actually force consumers to eat vaccinated meat? Can you ask this of the liberal consumer? When consumers come across vaccinated meat in the shops they don't buy it. The government must inform the public about the consumption of vaccinated meat.

4.3 Group 3

Opening up a debate within the EU framework on legislation governing control of FMD (Figure 4).

In order to create an international lobby, collaboration between the different interest groups (farmers' groups, veterinarians, etc.) needs to be established at European level, thus not only at policy and trade level. Policymaking at European level is not perceived as being democratic. This is why policymaking at this higher level should be done through interaction. There should be more insight into policymaking processes in Brussels. Research needs to be done on alternative structures for political decisionmaking processes. Furthermore, there is a need to gain insight into the cultural differences that influence decisionmaking in the member states and the new entries to the EU. These new entrants constitute a major veterinary problem. At the moment they form a good buffer, but once they have joined the EU this buffer will be inside the boundaries.

It should be possible to make differentiations in the legislation, for instance, different types of interventions per country, dependent on the destination of the export products.

Research problem	Expertise	Trajectory/Conditions
1. Tie in FMD policy with operating styles at European level (incl. risk analysis)	Beta-gamma integration livestock farmers, veterinarians, economists, sociologists, process supervisors, etc. FMA, WUR, rural sociology, ID/RIVM risk analysis	European approach
2. Alternatives to political decisionmaking trajectory in Brussels (incl. insight into what's happening now)	Beta-gamma integration cattle farmers, veterinarians, economists, sociologists, process supervisors, etc. FMA, WUR, rural sociology, ID/RIVM risk analysis	Coordinator LEI + Rathenau Institute + research institutes in other countries
3. Insight into cultural differences in political decisionmaking in the member states (plus new entries) topic: animal diseases in general	Management experts, politicologists, APS-ers, anthropologists LEI, ISS, KUN	Also involve consumers + other food chain parties
4. Trading policy considerations in the member states (incl. figures) as basis for FMD policy (Schengen as precondition)	Economists, management experts, veterinarians, expertise unit society	
5. EU contribution to development of animal husbandry in third countries to prevent FMD	Veterinarians, livestock farmers, beta-gamma integration	Coordinator: WUR/APS

Figure 4. Matrix group 3.

The open borders between European member states is laid down in the Schengen accord. But when problems arise the borders are closed immediately. This is viewed as contradictory. The borders should remain open and an outbreak of an animal disease is not a national problem but a European one. Insight into transport flows and the accompanying figures is needed. These are fundamental to the FMD policy. The policy should be dependent on the export destination, which can therefore differ per country.

There is a plea for research to be done at macro level (Europe) and micro level (the individual farmer) through which Netherlands policy should emerge. Risk analysis should be done of the different operating styles at European level as preparatory research for policymaking.

Europe must support the rest of the world in the veterinary field to reduce the pressure of infection.

Comments on the presentation

The European dimension is important for the risk analysis, among other things, Business Economics Group (FMA) ought to be involved in it.

In the presentation it was said that the risk of an outbreak of FMD is less in sustainable agriculture, but this is not the case.

5. First reaction of the Executive Board by Aalt Dijkhuizen

It was good that this took place – most definitely a WUR task. It would be a shame if WUR did not make use of it. It is important that WUR does something with it internally, although it is not that easy to get all the WUR groups to get involved in it. WUR must put up a united front when facing the outside world because an expression that is not univocal is not taken seriously.

The contents of the FMD scenario were technically and technologically perfect, the disease disappeared quickly and yet everyone is not satisfied. First, because the human factor was not taken into account. When an organisation is suddenly called into action people want to have a say in it. Second, little attention was focused on the process side. This has to do with the final result when difficult decisions have to be made. The role of the decision, the consensus and the considerations need to be made clear to the stakeholders to get acceptance of the measures that must be taken. Communication is essential here. This is political pluralism and demands a regional approach, but it does affect others. Therefore, research into the different regions in the EU and the impact on others is essential.

There are also other matters that were not aired such as who will pay for this research and what will the accompanying processes be and any untoward effects? Can the costs of a social cost-benefit analysis of the actors be disregarded? What are the preconditions and what is the context? Are we talking about the current situation (short term) or the longer term? There are many variables that play a significant role in the long-term, such as structure, sales and a new approach in the food chain.

What's next after this workshop? The results will need to be converted into communicable ideas, clustered together in main questions. There will need to be collaboration between different groups to execute this and support and financing will need to be sought jointly to prevent it from falling apart to be able to face the outside world as one. The best expertise must be made use of, which may also be brought in from outside WUR or from abroad.

WUR will not be the coordinator of the financing. This will need to come from different sources and the different clusters of research questions will need to make new arrangements for funding demands. There will be more involvement if financing comes from more stakeholders. The stakeholders willl need to present the plans properly.

This FMD debate is a topic with which WUR can profile itself. We should therefore take it up and into the future and not let it silt up.

Appendix 4. Participants in various activities

Persons interviewed

1.	ABE/FMA	M. Mourits
2.		F. Tomassen
3.		P. Berentsen
4.		M. Meuwissen
5.	ID-DLO	M. de Jong
6.		R. Schrijver
7.	CIDC	A. Dekker
8.	Dierenbescherming:	B. van den Berg
9.	Natuurmonumenten	T. Bosman
10.	Faculteit Diergeneeskunde:	A. Bouma
11.		M. Nielen
12.		E. Noordhuizen-Stassen
13.		J. Noordhuizen
14.	LLTB	H. Hermans
		J. Koolen
15.	LTO-Nederland	T. Cuijpers
16.	Expertisecentrum Ministry of Agriculture, Nature Conservation and Food Quality	E. van Klink
17.	CBL	W. de Jong
18.	Campina	J.F. Kleibeuker
19.		A.K. Schaap
20.	Alterra	P. Kersten
		H. van Haaften
21.	Ministry of Agriculture, Nature Conservation and Food Quality-VVA	S. Wiessenhaan
		A. Nielen
		J.E. Erkelens
22.	RVV	D.L. Schumer
23.	NLTO	A. Lanting
24.		K.J. Osinga
25.	Productschap Zuivel	F. Beekman
26.	LEI	T. Vogelzang
27.	GLTO	J. Roemaat
		M. S. Koopmans
28.	KNMvD	S.J. de Groot
		T. Jorna
29.	Actiecomité FMD	E.J. Aalpoel
30.	Rabobank	W.J.M. Thus
31.	Gezondheidsdienst voor Dieren	P. Vellema
32.	Rathenau Instituut	J. Staman
		L. Sterrenberg
33.	D66	P. Ter Veer
		F. van der Schans

34.	T.H.D.A	Z. Schaap
35.	Maatschappelijk werk	R. Bouwhuis
36.	Stichting Zeldzame Huisdieren	L. Elving

Both for Agriculture Economics Institute stakeholder analysis and the problem inventory for the FMD Workshop.

Participants in workshop 1: Stakeholders

Name:	Organisation:
Mr Th. Bosgoed	PMOV boeren
Mr. A.W.J. Bosman*	Vereniging Natuurmonumenten
Ms M. Bouwhuis-Tiggeloven	Maatschappelijk Werk Salland
Ms H. Bredenoord*	Ent Europa
Drs. L. Eland*	Gemeente Epe
Ms M. v.d. Engel	Ent Europa
Drs. J.E. Erkelens	Ministry of Agriculture, Nature Conservation and Food Quality-VVA
Ms J. van Eijk	Ned.Belangenver.Hobbydierhouders
Drs. S.J. de Groot*	KNMvD
Mr H. Hermans	LTB
Ms J. Hesterman	Ned. Belangenver. Hobbydierhouders
Mr B. Hopman	Nederlandse Vakbond Varkenshouders
Mr J. Klaver	Productschap Vlees, Vee en Eieren
Mr J. van Klink	Expertise Centre, Ministry of Agriculture, Nature Conservation and Food Quality
Mr D. Koelega	Ministry of Agriculture, Nature Conservation and Food Quality-Bsb
Mr A. Kon	NederlandseVakbond Varkenshouders
Mr. J. Laarman	GLTO
Ms. W. Mulder-v.d. Beek	Ent Europa
Ms. A. van Niersen-Vorselman	Werkgroep Landbouw en Armoede
Mr. F. van der Schans	D66
Ms R. Schut-Hakvoort*	Working Group, Agriculture and Poverty
Ms. J. Visser-Veldhuisen	Dutch Union of Dairy Producers
Ms. S. Wiessenhaan*	Ministry of Agriculture, Nature Conservation and Food Quality-VVA

Workshop Chairman
Mr H. Kieft

Presentation	
Ms.K. Eilers	WUR, DPS

Working group Chairpersons	
Ms I. de Boer	WUR, DPS
Mr B. ten Hoope	WUR, BC
Mr S. Oosting	WUR, DPS
Mr H. Sengers	Agriculture Economics Insititute
Ms A. van der Zijpp	WUR, DPS

Reporters
Mr W. Boer
Ms M. Braker
Ms K. van 't Hooft   ETC
Mr K. de Vries

Organisation
Ms K. Eilers   WUR, DPS
Ms M. Braker   WUR, DPS
*Representatives of the stakeholders workshop who reviewed the draft report.

Participants in workshop 2: Researchers

Name	Organisation
Ms M. Berendsen	Ministry of Agriculture, Nature Conservation and Food Quality-VVA
Ms G. van der Berg	Ent Europa
Mr A. Bianchi	CIDC-Lelystad
Mr A. Bosgoed	Larenstein, Deventer
Ms A. Bouma	University of Utrecht, FD
Ms H. Bredenoord	Ent Europa
Ms M. Commandeur	WUR, Agrarische Sociologie
Ms P.L. Eblé	CIDC-Lelystad
Ms I. Eijck	Praktijkonderzoek Veehouderij
Ms E.H. van Haaften	WUR, Alterra
Ms K. van 't Hooft	ETC
Mr P.H. Kersten	WUR, Alterra
Ms A. de Koeijer	WUR/ID-Lelystad
Ms M. Mourits	WUR, Agrarische Bedrijfseconomie
Ms A. Nielen	Ministry of Agriculture, Nature Conservation and Food Quality-VVA
Ms M. Nielen	University of Utrecht, FD
Ms K. Orsel	University of Utrecht, FD
Ms R. Oude Luttikhuis	Larenstein, Deventer
Mr. W. van der Poel	RIVM
Mr V. Pompe	Van Hall Institute
Ms A.M. de Roda Husman	RIVM
Mr D.L. Schumer	RVV
Ms R. Schut	Working group Agriculture and Poverty
Mr H. Sengers	Agriculture Economics Institute
Mr T. van der Ziel	WUR, Agrarian Sociology

Chairman	
Mr H. Kieft	ETC

Presentation:	
Ms A. v.d. Zijpp	WUR, DPS

Working group chairpersons	
Ms I. de Boer	WUR, DPS
Mr S. Oosting	WUR, DPS
Mr R. Schrijver	ID Lelystad
Mr T. Vogelzang	Agriculture Economics Institute

Reporters
Mr W. Boer
Ms M. Braker
Ms M. Kiep
Mr. K. de Vries

Organisation
Ms K. Eilers   WUR, DPS
Ms M. Braker   WUR, DPS

Members of the sounding board group of LEI-ABE-ID Lelystad

Mr V. Zachariasse	Agriculture Economics Institute
Mr G. Koopstra	Ministry of Agriculture, Nature Conservation and Food Quality-VVA
Mr E.J. Aalpoel	Actiecomité FMD-driehoek
Mr S.J. Schenk	LTO-Nederland
Mr A. Aalbers	NZO
Mr F. Stortelder	Dumeco
Ms H. van Veen	Dierenbescherming
Ms J. Pijl	Stichting het Gelders Landschap
Mr Tj. Jorna	KNMvD
Mr F. Pluimers	Ministry of Agriculture, Nature Conservation and Food Quality-VVA

The EAAP Technical Series so far contains the following publications:

- No. 1. Protein feed for animal production
  With special reference to Central and Eastern Europe
  edited by C. Février, A. Aumaitre, F. Habe, T. Vares, M. Zjalic
  ISBN 9076998035 – 2001 – 184 pages – € 35 – US$ 39

- No. 2. Livestock breeding and service organisations
  With special reference to CEE countries
  edited by J. Boyazoglu, J. Hodges, M. Zjalic, P. Rafai
  ISBN 9076998043 – 2002 – 75 pages – € 25 – US$ 30

- No. 3. Livestock Farming Systems in Central and Eastern Europe
  edited by A. Gibon, S. Mihina
  ISBN 9076998299 – 2003 – 264 pages – € 39 – US$ 51

- No. 4. Image of the Cattle Sector and its Products
  Role of Breeders Association
  ISBN 9076998337 – 2003 – 88 pages – € 27 – US$ 32

- No. 5. Foot and Mouth Disease
  New values, innovative research agenda's and policies
  A.J. van der Zijpp, M.J.E. Braker, C.H.A.M. Eilers, H. Kieft, T.A. Vogelzang and
  S.J. Oosting
  ISBN 9076998272 – 2004 – 80 pages – € 25 – US$ 30

- No. 6. Working animals in agriculture and transport
  A collection of some current research and development observations
  edited by R.A. Pearson, P. Lhoste, M. Saastamoinen, W. Martin-Rosset
  ISBN 9076998256 – 2003 – 208 pages – € 40 – US$ 53

- No. 7. Interactions between climate and animal production
  edited by N. Lacetera, U. Bernabucci, H.H. Khalifa, B. Ronchi and A. Nardone
  ISBN 9076998264 – 2003 – 128 pages – € 35 – US$ 39

These publications are available at:
Wageningen Academic Publishers
P.O. Box 220
6700 AE Wageningen   sales@WageningenAcademic.com
The Netherlands   www.WageningenAcademic.com

Wageningen Academic
P u b l i s h e r s

Printed in the United States
by Baker & Taylor Publisher Services